IMAGES
of America

DEATH VALLEY
GOLD RUSH

In 1992, the author took a trip to Death Valley with a friend who did not know how to get the spare tire off of his new jeep. On the dirt West Side Road, a tire blew, and the unfortunate choice was made to walk across the valley to the paved road on the east side. The trek included quicksand, desert pavement with flying stinging particles, sharp shards of salt rock, thick mud, delirium, and a forced overnight to rest. It took more than 10 hours to go eight miles while consuming more than three gallons of water each. Finally, the paved road was in sight, and there was a happy ending. Others in Death Valley's history had not been so fortunate. The stern lecture from the National Park Service was clear: "Never leave the vehicle. Someone will come." (Ted Faye.)

ON THE COVER: In 1905, mining investor and photographer Azariah Y. Pearl documented his trip to Death Valley with Walter E. Scott, known as "Death Valley Scotty," to confirm the value of Scott's claims (or if they existed at all). Scott's brother Bill is shown with two other scouts he hired. (California State Library.)

IMAGES
of America

DEATH VALLEY GOLD RUSH

Ted Faye

ARCADIA
PUBLISHING

Published by Arcadia Publishing
Charleston, South Carolina

Printed in the United States of America

Library of Congress Control Number: 2022936488

For all general information, please contact Arcadia Publishing:
Telephone 843-853-2070
Fax 843-853-0044
E-mail sales@arcadiapublishing.com
For customer service and orders:
Toll-Free 1-888-313-2665

Visit us on the Internet at www.arcadiapublishing.com

*To the prospectors, their burros, and to all of us who know
that what we're looking for "is just over that next hill"*

CONTENTS

ACKNOWLEDGMENTS

After more than 30 years of digging through nearly every archive I could find on Death Valley, I'm still amazed that fascinating new images continue to surface. I am most grateful to those private collectors who are willing to share, like Telly Eliades (TEC) who, along with his wife, Caroline, is dedicated to preserving history. "Bullfrog" Bill Miller (WMC) shared with me his passion for collecting all things Bullfrog and Rhyolite. Jim German (JG) is the great-grandnephew of pioneer Lorenzo Dow Stephens. There are longtime friends like Preston Chiaro, Henry Golas, Bobby Tanner, Brian Brown, Susan Sorrells, Joni Eastley, and Stanley Paher, who have been with me on this Death Valley journey since the beginning and whose support and encouragement have kept me going through the years. The work of extremely talented photographer Merilee Mitchell (MM) has made all of my projects better, and I am grateful for her contributions. Joan Patricia Breyfogle Bartlett (JPB) shared the only known photograph of her great-great-uncle Charles Breyfogle. Marv Jensen (MJ) and the Death Valley '49ers have helped me tell Death Valley stories for a wide variety of projects. At Death Valley National Park, I have appreciated the support of Supt. Mike Reynolds; curators and archivists Gretchen Voelk, Anne Llewellyn, and Greg Cox; and former longtime archivist Blair Davenport.

I found some new friends at museums who have gone the extra mile in helping me locate images, including Arlene Melton at the Central Nevada Historical Society (CNHS); Elaine Wiley at the Maturango Museum (MAT); Rachel Hads of the Kern County Museum (KCM); Heather Todd at the Eastern California Museum (ECM); Catherine Magee, director of the Nevada Historical Society (NHS); Laura Misajet at the Mojave Desert Heritage & Cultural Association (MDHCA); Julie Hoogland at the Palm Springs Art Museum (PSA); Marla Novo at the Santa Cruz Museum of Art and History (SCAH); Nicholas Piontek at the Center for Sacramento History (CSH); Dan Daniels at the Utah State University Special Collections (USSC); Kenzie McPhie at the University of Wyoming Frank Crampton Collection (FCC); the Jim Butler family (JBF); and Erin Chase at the Huntington Library and Museum of San Marino (HLM). The book also includes images from the Merced Historical Society (MHS); Brigham Young University (BYU); the Library of Congress (LOC); the California Historical Society (CHS); the Pomona Public Library and Henry Golas for Frashers Fotos (FF); and the California State Library (CSL).

Finally, thanks to some of my Death Valley friends who have passed away but had been primary consultants and offered support over the years, and whose research and work I still rely on. Richard Lingenfelter was the first Death Valley author I contacted when I knew I wanted to produce a documentary film on the subject. His definitive book on Death Valley history provides information included in this book. Sally Zanjani's personal connection to the Central Nevada gold rush was evident in all of her work, and her research is also reflected in this book. Suzy and Riley McCoy took care of the ghost town of Rhyolite for many years, and they helped by providing archival material and wonderful tales.

The author's collection is noted with (TF), and unless otherwise specified with the abbreviations noted here for other collections, the images presented in this book are from the Death Valley National Park Archives.

INTRODUCTION

It does not make sense that anyone would look at the godforsaken expanse called Death Valley and think, "I'm gonna get rich out there!" But that is precisely what happened over and over again from the mid-19th century until the 1920s. It is hard to fault anyone for thinking they could do it, as it had long been touted as a treasure trove of the world. People who had seen great ledges of gold and silver told their stories to the newspapers, and would-be prospectors read with great interest and went out to seek their fortunes. Native Americans looked on with curiosity, of course, as men with carts, shovels, picks, blankets, and burros began scouring their ancestral homelands for what they dubbed "color" rock—the color was shiny gold or silver.

In 1849, as the world was rushing to California, hundreds of thousands of prospectors earned the nickname "forty-niners" as part of that crowd, but none of them were purposely rushing to the deepest, driest desert in the Western Hemisphere. The destination was typically San Francisco or Sacramento, where these new arrivals could procure plenty of water, food, and supplies to go scoop up the gold nuggets awaiting their arrival. But this was a long journey by sea or land. From New York to San Francisco, going around the horn in a ship meant a 17,000-mile trip that took more than five months. Crossing overland in a covered wagon, the journey took six months or longer to travel the 2,000 miles from Missouri to California—plus up to three additional months depending on the origin point back east.

It was no wonder that those pioneers often sought shortcuts. However, taking one did not work out so well for the Donner Party in 1846, when a dreadful snowstorm stranded them in the Sierra Nevada Mountains. Some of the party resorted to cannibalism for survival, and the mention of the Donners sent shivers down the spine of anyone heading west in 1849.

So, when a group of westward-bound gold-seekers met in Salt Lake City in the fall of 1849, they decided not to head directly west on what was called the California Trail. If they had, by the time they reached the pass over the Sierra Nevada, it would be October—the same time of year when the Donners had met their doom. No one wanted that.

There was a trail to the south of Salt Lake City called the Old Spanish Trail. It had been used since 1829 as a pack mule route between Santa Fe and Los Angeles. Instead of heading due west, it made a big loop into Colorado and Utah, then went south through what would become Nevada and California to the pueblo of Los Angeles. In Utah, it reached north near present-day Provo, and it was there that Mormon guide Capt. Jefferson Hunt would lead a wagon train of more than 100 wagons, between 400 and 500 people, and more than 1,000 oxen, cattle, and pack animals. Only one wagon had been on the route; it was poorly watered, there were stretches of harsh desert, and it was 500 miles longer than the California Trail—but it was better than resorting to cannibalism.

Stragglers from this group tried to take a shortcut and ended up in the desert trench they named Death Valley. It is likely that had they named it anything else, such as Happy Valley (a suggestion made by the author's father after he enjoyed a wonderful visit there), in all probability, this book—or any other book on its extraordinary history—would have never been written. But according to pioneer forty-niner William Lewis Manly, once they named it that "saddest and most dreadful name," the course of its history was set.

Within the first year of their escape from the valley named for death, some members of the group returned to look for ledges of silver and chunks of gold and to try and retrieve treasure they had buried on the trail. Over time, the legends grew and attracted more prospectors and others like Charles Breyfogle, who found and lost a great mountain of gold, bringing others back with him in an effort to find it. Though this group was unsuccessful in finding the lost treasure, in the process of looking for it, some members made a rich strike and claimed it was old Breyfogle's, the Lost Gunsight Silver, the Lost Goler, or some other lost mine. The rumors and finds kept the mystery and lore of Death Valley percolating for more than 50 years.

In the 1890s, the forty-niners had reunions and recounted their experiences in books and newspapers. Pacific Coast Borax began using Death Valley to promote its 20 Mule Team Borax products. An official federal government expedition had been sent to explore this mysterious valley, and in 1894, a writer for the *New York World* described the land as the "pit of horrors, the haunt of all that is grim and ghoulish." Death Valley was gaining a reputation as the wildest place in America and a tortuous and tormented landscape guarding a mass of riches. Then, in 1900, about 100 miles northeast of the dreaded valley, an event occurred that seemed to make all of Death Valley's rumors and lore a reality.

Jim and Belle Butler lived on a ranch near the Nevada town of Belmont. Carl Glasscock, in his book *Gold in Them Hills*, noted that Butler was known as "Big Jim" to his friends and "Lazy Jim" to his enemies. He told great stories but often lost enthusiasm for projects he started. At heart, Jim was a prospector, and one day, Jim rested with his burros in the San Antonio Mountains near a spring the Native Americans called Tonopah. Though there are varying accounts, Glasscock records this story of Tonopah's discovery. On May 19, 1900, Jim's burros went wandering, and Jim went looking for them. It was a windy day, and "the burros had found shelter behind a ledge, where Jim discovered them. Appreciating their wisdom he sat down to wait out the wind, and as he sat, chipped idly at the rock. The specimens that came away in his hand appeared sufficiently promising to carry away."

Jim had trouble getting the rocks assayed right away, but once he did, they were found to be rich in silver and gold. His wife, Belle, badgered him enough that they both went and formally staked the claims at Tonopah. She, by the way, struck the biggest claim, calling it the Mizpah, invoking a biblical blessing. Glasscock mused that "one lazy lucky prospector had broken the slump which had been depressing the West." Not since the glory days of Virginia City and the famed Comstock Lode had Nevada seen such riches uncovered.

Word spread, luring prospectors, promoters, businessmen, and merchants from all over the world. The Klondike gold rush to Alaska had just ended, and Tonopah was the newest and richest discovery on the Western frontier. With the fledgling desert town as a home base, prospectors began scouring the nearby mountains, gullies, and ledges in search of more riches.

Just two years after the strike at Tonopah, another rich discovery was made about 20 miles to the south. Glasscock noted that "how prospectors get their information, across scores and hundreds of miles of desert mountains and valleys, without roads, without radio, without wires or mail, probably will remain a mystery." In this case, it was no mystery. Prospectors William Marsh and Harry Stimler were financed by Tonopah businessmen (including Jim Butler) to locate the source of a valuable gold specimen brought to them by Shoshone prospector Tom Fisherman. Of course, as happened to many Native Americans, they were never compensated for their knowledge of the location of valuable minerals. When Marsh and Stimler located the rich claim, they believed it was the granddaddy of all claims, so they called it Grandpa. This area later became known as Goldfield.

From 1902 to 1903, Goldfield's population grew from around 50 hopeful souls to more than 5,000 fortune-seekers. Glasscock said that "so rich was its ore that one newspaper announced that [Goldfield] must be the famous lost Breyfogle—perhaps the most famous of all the great lost mines of the desert." Goldfield sprang up literally in the middle of nowhere, not even near a good spring like Tonopah. It had substantial buildings made of wood and stone. Plans were made for Goldfield to last for a long time.

By 1904, nearly $2.5 million in gold and silver had been produced, and the world was taking notice. The *London Financial News* feared that gold from the mines might flood the market, thus devaluing money and creating a financial panic. That never materialized, but what did happen in the summer of 1904 might have given hope to many that the strikes would go on forever. This is the point where Death Valley makes its entrance.

In August 1904, on the edge of Death Valley, prospectors Frank "Shorty" Harris and Ed Cross found gold in some green rock about the shape, size, and color of a bullfrog. The Bullfrog claim had been struck, and another rush was on. Death Valley's time had come. There had been rumors, tales, and legends of gold and riches in the dreaded valley named Death, and here it was. No one could deny it, and throngs of hopefuls began heading to the most feared location in America.

Shorty Harris was the embodiment of what came to be known as the single blanket jackass prospector. With just a blanket, a jackass (burro or donkey), a shovel, a pick, bacon, and beans, this type of prospector was ready to go out and look for gold. It is worth noting that many make the mistake of confusing a donkey and a mule. A jackass, burro, and donkey are the same. A mule is the spawn of a male donkey and a female horse—a hybrid animal historically used for packing, riding, and pulling. Images of Harris always show him with a jackass.

Harris was born in Rhode Island in 1857, headed west in the 1870s, and then prospected in locations ranging from Colorado to Idaho and even Tombstone, Arizona—the site of Wyatt Earp's infamous gunfight at the O.K. Corral. But in all of his adventures, nothing matched the frenzy he and his prospecting buddy Ed Cross started at Bullfrog.

"I've seen some gold rushes in my time," Shorty Harris recalled to Philip Johnston in 1930, writing for *Touring Topics*, "but nothing like that stampede. Men were leaving town in a steady stream with buckboards, buggies, wagons, and burros. It looked like the whole population of Goldfield was trying to move at once. They all got the fever and milled around wild-eyed, trying to find a way to get to the new 'strike.'"

Harris's personal story is typical of so many prospectors. He sold out too soon and for too little. Going on what came to be called the Grand Spree, Harris hit the Goldfield saloons hard and woke up six days later in a drunken stupor having sold his interest in the claim for $1,000. Cross had been trying to line up a sale for $10,000 but could not find Harris. However, he did find the man Harris sold to and made a deal, eventually selling his part for more than $125,000 (nearly $4 million in today's money). With his paltry $1,000, Harris bought more drinks for his friends.

Though the bustling and optimistic town of Bullfrog sprang up, it soon gave way to what would become the queen city of the Death Valley region: Rhyolite. Named for the type of volcanic rock found in the area, Rhyolite owed its existence, in large part, to Bob Montgomery, who was familiar with the area when he and his brothers made a strike in the Panamint Mountains on the west side of Death Valley in the 1890s. Bob came for the Bullfrog rush and found nothing. But on his way back to Goldfield, he met a young Shoshone named Hungry Johnny and hired him to check out some possible claims in exchange for grub (food) and a promise for more, should he stake a claim. This process of hiring a prospector to look for gold was called grubstaking.

When Montgomery returned some weeks later, Johnny showed him some likely pay dirt, so Bob staked claims for about a mile in the region. Though Montgomery came back several times, it was not until an old prospector showed him precisely where to dig on his claims that he was rewarded. In 1905, Montgomery ran a tunnel that struck bonanza ore running 70 feet thick with assays valued at $16,000 per ton. In his book *Death Valley and the Amargosa*, Richard Lingenfelter quotes Montgomery as saying: "I have struck it; the thing that I have dreamed about since I was 15 years old has come true; I am fixed for life and nobody can take it away from me!" Montgomery's mine was called "the new wonder of the west" and hailed as one of the greatest discoveries ever made. Death Valley was giving up its riches.

The year 1905 was a big time for Death Valley. This was when it became famous for its gold, borax, colorful characters, mule teams, and dangers. This was also Death Valley's deadliest year to date. Young prospectors from all over the world, unfamiliar with the desiccating heat and dryness of Death Valley, often found themselves in life-threatening situations, becoming dehydrated, confused, and delirious, with bloody cracks appearing as their skin lost all of its moisture. Reportedly, some desperately removed all their clothing, perishing while screaming how deep the water was around them.

The entire Death Valley rush was fueled by investors like Charles Schwab and promoters like George Graham Rice. Schwab, as president of Pennsylvania's Bethlehem Steel, saw the value of mineral investment, and Rice saw the value of selling worthless mining stocks. Characters like Death Valley Scotty benefitted from both, as he touted his secret Death Valley gold mine to secure money from the pockets of potential financiers.

Staying largely out of the boom excitement was the Pacific Coast Borax Company. In 1906, the company returned to open a mine in the mountains on Death Valley's eastern side. Borax represented the corporate and industrial interests, though it was not above using Death Valley's mystery, lore, and excitement to help sell 20 Mule Team Borax products. Though longline teams had not hauled borax from Death Valley for nearly 20 years, it did not stop the company from taking its historical wagons on a nationwide tour to tout the borax that came straight from Death Valley.

Eventually, Rhyolite and the mines that resulted from the boom—like Skidoo, Greenwater, and Keane Wonder—all faded out, and by 1910, most everyone was gone. The era of Death Valley's ghost towns had arrived. The gold rush was gone, but a new rush began. Tourists were curious as to how quickly people had left. Dishes were left on the tables, checks were left in the bank, and a printing press was discarded in the desert. The once-thriving hubs of desert life had become ghostly tourist attractions.

So, it all came full circle. The grand desolation described by pioneer William Lewis Manly would become the reason people would visit. A new kind of gold rush—in tourism—began to take hold in the 1920s. As the last flurry of old-time mining stock promotion took place, staged by con man C.C. Julian, plans were also being made to incorporate Death Valley into the National Park System.

From the heady days of 1849 into the 1920s, Death Valley gained its reputation and forged its history. Today, the reasons for visiting this unique and famous valley have changed from seeking mineral wealth to looking for scenic beauty. There can yet be found the remains of the dreams of riches that fueled the Death Valley gold rush.

One

THE JAWS OF HELL

In October 1849, a group of pioneers heading west for the California gold rush met at the tiny Mormon settlement near Utah's Salt Lake. From there, the most direct path to the gold country was due west, but it was late in the year, and winter was coming. The pioneers heard horrific tales of the Donner Party—just three years earlier, they had been stranded in the mountain snow and resorted to cannibalism.

An alternate route to the south called the Old Spanish Trail had been traveled by pack mules but not wagons. Mormon guide Capt. Jefferson Hunt offered to guide the wagon train on the trail southwest through Utah, over the Basin and Range portion of Nevada, into the Mojave Desert of southern Nevada and California, and ultimately to the pueblo of Los Angeles. From there, they would venture up the coast to Sacramento.

The group left Salt Lake and gathered at Fort Utah (present-day Provo) to head south and meet the Old Spanish Trail. Though it saved them from the winter snows, the trail was months longer than the direct route. It was no wonder that the pioneers got excited over a map with a shortcut. Twenty-five wagons turned west, heading straight across the Nevada desert. Hunt warned that taking the shortcut would lead them "into the jaws of hell." They ignored his counsel, splintering into smaller groups as they wandered aimlessly around mountains and over dry lake beds, ultimately meandering into the lowest, driest, hottest place on the planet.

Most were fortunate to make it out alive. Though they had abandoned all their earthly possessions and escaped with only the clothes on their back, their horrific memories would define their lives and the valley forever. On the trail, one man found a black rock rich in silver, and a Mormon missionary in another group found chunks of gold. Others suffered terribly and christened the place Death Valley. It was not long before rumors abounded of the ominous and horror-filled area known as Death Valley, which was known to be guarding a treasure trove of riches.

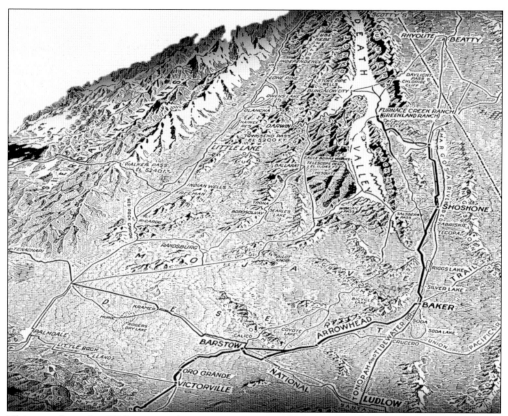

This map from the 1930s was used in many promotional tourism brochures. Los Angeles is to the lower left, and Las Vegas is to the right. Just south of Beatty, the forty-niners headed south along the Amargosa River (which flows mostly underground). To the left of the "r" in Amargosa, the pioneers entered Furnace Creek Wash, which took them into what Mormon guide Capt. Jefferson Hunt warned were "the jaws of hell." (TF.)

In 1849, the Timbisha-Shoshone tribe lived at the mouth of Furnace Creek Wash. They stayed in wickiups (huts made of brush), and to them, this was not a valley of death or riches—it was home. They lived in the valley during winter, and in summer, they climbed to cooler elevations. This 1920s photograph shows a Timbisha camp at Furnace Creek. Note that the traditional wickiups have been replaced with tents.

Native American George Hansen (seen here at his Indian Ranch west of Death Valley) lived to be 115 and gave a compelling eyewitness account of the events of the winter of 1849–1850. As a boy, he marveled at the carts being pulled by skinny cows. He ran away from the white men, giving rise to his nickname, "boy who runs away." The Timbisha knew the pioneers were invaders.

Panamint Tom Wilson was a young Native American boy when the gold-rush pioneers arrived. He had his own tales of riches when he talked about an ancient city of gold in Death Valley's underground that was visited by his grandfather. The story, later written by Bourke Lee in *Death Valley Men*, told of a fabulous ancient civilization of gold hidden underground in the Panamints.

In 1849, far from the desert valley later called "Death," gold-seekers panned nuggets from the rivers of northern California. Sacramento and San Francisco boomed as thousands rushed in from all over the world. In 1848, when gold was found in a stream near Sutter's Mill, word got out, and the rush began. These 1850 hopefuls posed with their pick and shovel on the way to the diggings. (CHS.)

In 1849, the reality of mining life was harsh. This image of gold-rush prospectors in El Dorado, California, illustrates the difficulty of the work. Flumes used to dredge out the hillsides were constructed to unearth the gold. But it was mostly hours, days, weeks, and months of endless digging with very few people acquiring the riches they sought. (LOC.)

This illustration from the time of the gold rush depicts a prospecting party headed to the goldfields. As they start their journey, they are in good shape and energetic, and by the time they return, they have lost their mule, their clothes are shredded, and they have no gold. (LOC.)

During the gold rush, the port of San Francisco became so congested with ships coming from all over the world that there was often no docking space. Eager prospectors would jump ship in the harbor to get a smaller boat or swim ashore to get to the goldfields as fast as they could. This image from 1851 shows the congestion at the port. (CHS.)

In 1849, a train of wagons assembled near Salt Lake. It consisted of about 110 wagons, 400 pioneers, and 1,000 animals (oxen, cattle, and mules). The members of the train called themselves the San Joaquin Company but were nicknamed the Sand Walking Company. They arrived in the fall and feared crossing the Sierra Nevadas, where winter snows might block their progress and doom them to the fate of the Donner Party, some members of which had resorted to cannibalism. (LOC.)

This 1870s photograph of Salt Lake City features the tabernacle. In 1849, the structure was much more primitive, with only poles and a roof of sagebrush. It sat about 1,000 people. Pioneer Lorenzo Dow Stephens recalled, "I have been asked how many wives [Brigham Young] had and all I have to judge by was the sleeping apartments. I saw seven or eight wagons with covers on, all in a row . . . and was told these were occupied by his wives." (BYU.)

The forty-niners left Fort Salt Lake and proceeded south to Fort Utah (Provo). There, Capt. Jefferson Hunt was chosen to be the group's guide. He said they would take the Old Spanish Trail, and it was his responsibility to lead them safely to California. It was not long before some accused him of having a faulty memory and being unable to find water or the exact route. (CSL.)

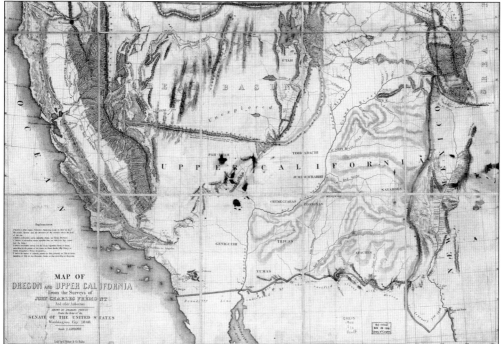

John Fremont's 1848 map shows an imaginary east–west mountain range. Pioneers who trusted the map imagined streams and rivers at the base of the range. Instead, what they encountered was some of the most treacherous terrain in the United States, a rugged desert in the heart of what is now called the Great Basin. It was this map that convinced many in the San Joaquin party to make a fateful mistake. (LOC.)

This map shows the primary trails west. The Old Spanish Trail begins in Santa Fe and heads north into Utah, then south through Nevada, and into California. Lorenzo Dow Stephens writes, "As soon as the wagons were in readiness the start was made on the first day of October. With such a large company we made but a few miles a day. The trail was over low rolling hills covered with scrub cedars and somewhat sandy soil in places." (TF.)

Soon, three Mormons joined the San Joaquin party and suggested a shortcut. Addison Pratt writes that Reverend Brier "took the opportunity to fire the minds of the people with zeal for the cutoff . . . by saying . . . sink or swim, live or die, he should take the cut off 'go it boots!'" On Sunday, November 4, 1849, nearly 100 wagons took the cutoff. This monument near Enterprise, Utah, marks the site of the cutoff. (TF.)

While most of the San Joaquin party's wagons returned to follow Capt. Jefferson Hunt, a group known as the Jayhawkers, from Illinois, kept to the shortcut. John Colton (above left) and Capt. Edward Doty (above right) led the Jayhawkers, though it was likely the 18-year-old Colton who organized the company of approximately 12 wagons and 35 men. In later years, following their ordeal, Colton organized the group's reunions.

At age 20, William Lewis Manly journeyed from Wisconsin, and John Rogers, 27, came from Tennessee. Manly (pictured) was respected for his scouting skills and knowledge of the land. He wrote in *Death Valley in '49*, "About three days journey on the new trail . . . in front of us was a canyon, impassible for wagons, and down into this the trail descended." The spot from which they looked into the canyon was dubbed Mount Misery.

Published in 1894, William Lewis Manly's book *Death Valley in '49* became a pioneer classic. He wrote about the desert mirage: "I . . . learned that when clear lakes appeared . . . we might search and search and never find them." He describes his loyalty to two families: "I felt I should be morally guilty of murder if I should forsake Mr. Bennett's wife and children and the family of Mr. Arcane [sic] with whom I had been thus far associated." (TF.)

Asabel Bennett was a longtime friend of William Lewis Manly, with both of them having left for the goldfields from Mineral Point, Wisconsin. Bennett was about 42 when he headed west with his wife, Sarah; their son George, who was then eight years old; and their daughters Melissa, who was five, and Martha. Asabel (left) and Martha are pictured here—Martha was only a year old on the journey but is shown here as a young lady.

Jean Baptiste Arcan was a French Basque born in 1813. His wife, Abigail, was Danish. Heading west, they brought their three-year-old boy, Charles. The Bennetts and the Arcans, along with John Rogers and William Lewis Manly, took the shortcut. They would all play a part in one of the West's most historic and legendary rescues, ultimately giving an unknown desert valley the name that would change it forever. (SCAH.)

Harry Wade was born in England in 1800 and became coachman to the king. In 1835, he brought his wife, Mary, to New York, and by April 1, 1849, they were heading west with their children, Charles, Almira, Richard, and Harry. They also took the shortcut, but only Harry and one other family safely got their animals and wagons out. Some resented Wade's perceived aloof attitude, but he was an experienced freighter intent on survival.

Methodist minister Rev. James Welch Brier and his wife, Juliette (Julia) left Michigan, took their three boys (Christopher Columbus, John Wells, and Kirke White), and headed west to be missionaries. John Colton wrote that after they abandoned their wagons in Death Valley, "Mrs. Brier carried one of the children in her arms, another on her back . . . and a third she led by the hand. There was never a murmur from this brave woman." (CHS.)

Lorenzo Dow Stephens grew up in Illinois, where, at the age of 21, he joined the Jayhawkers for the gold rush. He became an influential rancher, invested in lumber and railroads, and chased gold strikes. He wrote his memoirs in 1916, attended Jayhawker reunions, and died in 1921 in Oakland, California, at the age of 93. This is his 1867 wedding photograph with his wife, Julia. (JG.)

Lorenzo Dow Stephens was born here in 1827. The home, built in 1815, is typical of what the forty-niners left behind. The contrast to their Death Valley experience was stark, as William Lewis Manly writes, "I thought of the bread and beans upon my father's table, to say nothing about all the other good things, and here was I away out in the center of the Great American Desert, with an empty stomach and a dry and parched throat, perhaps I had not yet seen the worst of it." (JG.)

Nearly all who took the shortcut came through this narrow, rocky pathway. William Lewis Manly wrote, "It stood in sharp peaks and was of many colors. I imagined it to be a true volcanic point and had never been so near one before, and the most wonderful picture of grand desolation one could ever see." The peak was later named Manly Beacon, the mountains in the background were dubbed the Panamints, and the pathway was called Furnace Creek Wash.

The Wade family and several others continued south through the valley, taking their wagons and eventually finding a way out. The Bennett and Arcan families, however, along with William Lewis Manly and John Rogers, found a small spring and camped here about 20 miles from where they had first entered the valley at Furnace Creek Wash. This monument commemorates the Long Camp, where it was decided that the families would wait while Manly and Rogers went for supplies and help.

The "boys," as William Lewis Manly and John Rogers were called, left camp and headed into the desert on January 15, 1850. After 11 days of trailblazing, following Native American paths, and finding pastureland, they came to a small hacienda about 20 miles north of the San Fernando Mission (pictured). There, they purchased beans, wheat, meat, and oranges and were given three horses and a little one-eyed mule. On January 30, they headed back to rescue their friends. (ECM.)

24

LEAVING DEATH VALLEY,—THE MANLY PARTY ON THE MARCH AFTER LEAVING THEIR WAGONS.

As William Lewis Manly and John Rogers made their way back to the valley, the horses died, but the little one-eyed mule persisted. As they got closer to the camp, they found a body—Captain Culverwell, who was ill and traveling alone. He tried to walk out but never made it. As Manly and Rogers approached the wagons, they saw no signs of life, so they fired a shot into the air. From under the wagons came the families, shouting, "The boys have come!" (TF.)

The families abandoned their wagons. The women rode the oxen with the children in sacks. When the sacks shifted, one ox, Old Crump, tossed Sarah Bennett. They all had a good laugh, but the next day, as they crossed the Panamints, they looked back over the valley. William Lewis Manly writes that they "spoke the thought uppermost saying, 'Goodbye Death Valley . . . ours was the party which named it the saddest and most dreadful name.'" (TF.)

25

At "ox-jump falls," William Lewis Manly writes, Asabel Bennett and Jean Baptiste Arcan "urged one of the oxen to the edge of the falls, put the rope around his horns and threw down the end to me . . . I was told to pull hard . . . that he might not light on his head and break his neck. Those above gave a push and the ox came over, sprawling, but landed safely. The little mule gave a jump when they pushed her and lighted squarely on her feet." (TF.)

William Lewis Manly drew this crude map, writing, "You can see our different camps in Death Valley by round dots all along our trails from the dry lake to the Spanish Route. It no doubt will puzzle you to understand my crude work marked out after 45 years have rolled away." The Bennett and Arcan families, along with Manly and John Rogers, entered Death Valley on Christmas Eve in 1849 and made it to the ranch house near Los Angeles on March 7, 1850. (TF.)

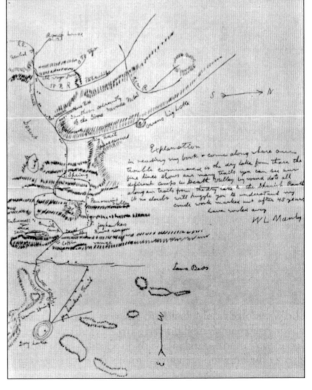

While William Lewis Manly, the Bennetts, and Arcans went south in the valley, the Jayhawkers headed north. At a place near today's Stovepipe Wells resort, they burned their wagons, taking only what they could carry out over the mountains and into the next valley, Panamint Valley. This monument at Stovepipe Wells commemorates Burned Wagons Point. (TF.)

For nearly a century after the forty-niners went through Death Valley, the remains of their wagons and possessions could be found. Most were sold as souvenirs, and other items went home with collectors. In this 1920s photograph, Ralph Jacobus Fairbanks (left; better known as "Dad") and Panamint Tom Wilson (Shoshone Native American) inspect the remains of the Jayhawkers' burnt wagons.

The sons of Capt. Ed Doty, Henry (center, holding the gun his father carried through Death Valley) and Frank (right), pictured with Henry's son Chester, stand at the site of the old Rancho San Francisquito near Castaic. It was here that the forty-niners were treated with hospitality by the Californios who sustained them after their ordeal. The gun survives in the collection of Death Valley National Park.

In 1994, the author was notified that a delicate tablecloth owned by Abigail Arcan still survived in the hands of Arcan descendants. Abigail did not want to leave her finest tablecloth behind to be gathered up by Native American raiders, nor did she want the family to lose their best clothes. They dressed up in their Sunday finest, and she wrapped this tablecloth around her waist to take it out of the valley. In this picture, the author (left) inspects the tablecloth while its owner, Ed Stark, looks on. (TF.)

On a corner of the tablecloth taken out of the valley by Abigail Arcan, an inscription attests to its provenance. Abigail's daughter-in-law Etta Arcan Ryan states that Abigail brought it "from Illinois across the plains and deserts and that awful Death Valley." The tablecloth was notarized in 1941. The Stark family eventually donated it to the Santa Cruz Museum of Art and History. (SCAH.)

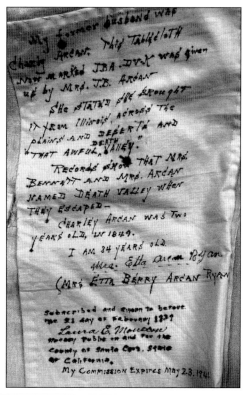

EVENING SUN

MERCED, CALIFORNIA, SATURDAY, DECEMBER 29, 1906.   No. 219

RIC CARS TO YOSEMITE

AD FROM MERCED ONCE BE CHANGED ELECTRICITY.

Bagby, Dec. 27, 1906.
ite Valley railroad is to into an up-to-date electric t as men and money can ontract for the poles, as e telegraph line, has been ullivan, formerly of Sulli- ds, of San Francisco. Mr. just completed a large he Valley Power Co., in and his outfit is now at d will be shipped herə

best news the Sun has ers for some time. There n, a something about an ay to the Yosemite that ing to the old steam same glistening, spark- at electrifies and charms s it leaps over the many osemite is harnessed, so n in the Merced canyon. er is used to carry the the wonders of the grea:

not mean the abandon team locomotive. It gives uble system. Should the shut down at any time oil-burning engines are ke the place. There will y special trains go over

AN OLD MERCED PIONEER JOINS THE MAJORITY.

Another of Merced's pioneers passed away last night when Jesus Aros succumbed suddenly to heart disease when quietly sitting in the Mint sa loon at about 12:30 o'clock. In the death of the man this city loses its most prominent figure of early sport- ing days. In the '70's Merced was not the modern municipality of today, and ated. The red lights burned late and the cowboys and miners eagerly look ed forward to days of boisterousness on their trips to the settlements. Mer- ced obtained its full share of that class. The resort conducted by Jesus Aros was then known far and wide and many of our old timers, had they the inclination, could recall occur- rences there which are not included in our various local published histories. Money was easily earned and easily spent, and the pop of guns was signi- ficant that the accepted law was being enforced. Aros was never in trouble, and despite his business he was gen- erally well thought of. His word was as good as his note. There were no vicious instincts in his make-up. With him his business was simply a legiti- mate means of earning a livelihood and fully up to the standard accepted on the frontier. He was courteous to all, an interesting conversationalist and a versatile musician of no mean capacity. He was also an active mem- ber of the old Eureka Hose Company and took considerable pride in relat- ing the accomplishments of that or- ganization. The deceased was a na- tive of Sonora, Mexico, and about 65 years old. The arrangements for the

JOHN ROGERS TIRES OF LIFE

ATTEMPTED SUICIDE FRIDAY AS A RESULT OF PHYSICAL IN- FIRMITIES.

Overburdened with an abundance of physical ailments, John Rogers, a pio- neer resident of this locality, slashed himself into ribbons with a razor in the county hospital yesterday in an at- tempt to end his troubles. He was found a few minutes afterward rapid- ly bleeding to death. His neck was gashed in six different places, though the blade had failed to touch a vital part. The desperate man had also inflicted three wounds upon each of his arms at the elbow joints. County Physician Castle was promptly sum- moned and after staunching the flow of blood stitched the wounds. Rogers is not expected to live, though today he seems to be gaining strength.

The unfortunate man is 84 years of age and was one of the first inhabi- tants of this locality. About a year ago he was admitted to the county hospital and many times since then he has threatened to take his life. Rogers is an old time miner and in his younger days lost a portion of his left leg because of mercurial poison- ing. A portion of his right foot was also amputated. The old man then lost one of his eyes, and to add to his troubles the sight of his other eye has of late been failing him. This fact preyed upon his mind and in conver-

though afterward he expressed sorro at his act. "Old man Rogers" known all over the county. It is sai that many years ago he was city ma shal of Gilroy.

Later—Rogers died about 1 o'cloc this afternoon.

SUNDAY SERVICES AT THE VARIOUS CHURCHE

Catholic Church.
Early mass, 8 a. m.; late mass 10:30 a. m.; catechism immediate afterwards. Rev. Father McNama will officiate.

St. Luke's Episcopal Church.
Sunday after Christmas, Decembe 30, 1906. Morning prayer and sermo read by Mr. A. R. Gurr, lay rector, 11 a. m. No service in the evening. Rev. Prof. N. Saunders, Rector.

First Presbyterian Church.
Services will be held tomorrow a follows: Sunday school at 9:45 a. m morning service 11 a. m.; evening se vice at 7:30 p. m. The public is co dially invited to each of these service George S. Greig, Pastor.

Cumberland Presbyterian Church.
Sunday school at 10 a. m., O. Baker, superintendent. Preaching 11 a. m. At this service two deaco will be ordained and the sermon wi be on "The office of Deacon.' Inte mediate C. E. Society at 3 p. m., wil Zoe Neal as leader. Junior C. E. S ciety at 3 p. m. Senior C. E. Socie at 6:30 p. m. with Amy Amsbaugh a leader. Preaching at 7:30 p. m. T these services the public will find Christian welcome.
James Miles Webb, Pastor.

The fate of John Rogers was unknown until the author and research associate Robert Ryan diligently looked for him. After locating a death certificate and finding his obituary, it was discovered that the end of such a hero was a sad one. Rogers went blind and committed suicide in an infirmary in 1906. While there are no known photographs of Rogers, he was as important to the rescue of the families as William Lewis Manly. (MHS.)

While the death certificate and the obituary found by the author and research associate Robert Ryan (the man at left) listed the cemetery where John Rogers was buried, there was no marker for his grave. With the help of the plot map and the good folks of Merced, in 1993, the exact location of Rogers's grave was found. In this image, Ryan observes prodding for the location. (TF.)

It was not until 2006 that the Death Valley '49ers, an organization dedicated to preserving the history of Death Valley, would place a marker for John Rogers. The Death Valley '49ers sponsored the effort, which was spearheaded by Marv Jensen, whose late wife, Mary, was related to the Wade family. Pictured here are, from left to right, Death Valley '49ers board members Bette and Ray Sisson, Bill and Edie Poole, and Jensen. (MJ.)

This monument is located at the base of Death Valley's Furnace Creek Wash. Placed in 1949 as part of a statewide California centennial celebration, it marks the 100th anniversary of the forty-niners' horrific experience. It is noteworthy that while many of the pioneers suffered greatly, the Wades and one other wagon made it out safely. Author William Lewis Manly never mentions that fact in his book, *Death Valley in '49.* Perhaps, as author Richard Lingenfelter suggests, "it would have spoiled his heroic rescue of the Bennetts and Arcans." (TF.)

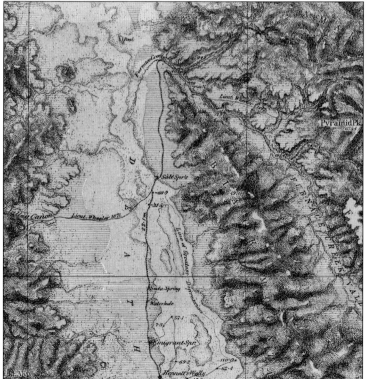

Lt. George Wheeler surveyed the Death Valley country in 1876, producing this map. On the right is Furnace Creek valley, and, at the mouth, the words "Furnace Creek." It was here that the Jayhawkers headed north and the Manly, Rogers, Bennett, Arcan, and Wade parties headed south. On the floor of the valley, along what is today's West Side Road, various watering holes are marked, including Bennett's Well. (TF.)

Once William Lewis Manly and John Rogers had brought their friends out of Death Valley, Manly writes, "I said to myself the only way to keep me from getting to the gold mines was to kill me. I felt that there was not a mountain so high I could not climb, and no desert so wide and dry that I could not cross it . . . I felt so strong and brave I could do it again—any way to reach the gold mines and get some of the 'dust.' I had not much idea how the gold from the mines looked. Everybody called it gold dust, and that conveyed an idea to me that it was fine as flour, but how to catch it I did not know. I knew other people found a way to get it and I knew I could learn if anybody could. It was a great longing that came to me to see some of the yellow dust in its native state, before it had been through the mint." None of the Death Valley forty-niners ever made their fortunes in California's goldfields. (PSA.)

Two

They Lost Their Mines

The pioneers who left Death Valley in 1850 took with them not only tales of horror, death, and survival, but also of silver, gold, and a desolate landscape full of treasure. Soon, those who barely escaped with their lives from the valley returned to seek the fortunes they believed were hidden there.

John B. Colton, one of the original Death Valley Jayhawkers, recalled in an 1895 article that, as they were escaping that horrific valley, "Captain Jim Martin, from the party of Georgians showed me a chunk of black rock . . . and told me it was half silver . . . and that there was all the wealth in sight that we could ask."

Martin was partly right, as there was indeed silver in the rock. When Martin reached either Los Angeles or the mines at Mariposa, he had the silver made into a gunsight. When word spread that Martin had made his discovery in Death Valley, legend grew of a great "Silver Mountain," and the quest for the "gunsight silver" began. When Colton wrote about the Lost Gunsight Mine in 1895, he lamented that "hundreds of lives have been lost in the hopeless search for 'Silver Mountain.' Allured by the stories of the fabulous wealth contained in the hidden mines of Death Valley . . . many an unfortunate prospector has left his bones to bleach in the wilderness."

In 1864, fourteen years after Martin's discovery, Charles Breyfogle was running a hotel at a mining camp near Austin, Nevada. When the mining camp folded, he took a party of four men and set out for Death Valley to find the source of a chunk of gold he possessed. He could never find the lost ledge of gold, but over the years, he kept trying, and though he could never produce any evidence, more and more prospectors believed his lost gold was out there. And so, the "Lost Breyfogle" became Death Valley's most famous unrecovered gold lode.

Over the years, many quests for the gunsight silver and the Lost Breyfogle opened Death Valley to exploration and other significant discoveries, as well as leading to tales of more lost mines.

While not a lost mine, this event is in keeping with the perpetuation of frauds. This image shows a young John Colton shortly after his arrival in San Francisco following his Death Valley experience, where he lost 100 pounds. The photographer peddled this photograph to curious miners as featuring a girl. According to Frank Latta in his book *Death Valley '49ers*, Colton himself exposed the fraud. (HLM.)

This 1903 article proclaims, "These were the Jayhawkers who participated in the discovery of Silver Mountain!" It was true that as Jayhawker Jim Martin struggled out of the valley, he did find a rock rich in silver, which he made into a gunsight. The Lost Gunsight silver legend was born and became known throughout the West. How it evolved into the tale of Silver Mountain is one of Death Valley's enduring mysteries.

William B. Rood was one of the Jayhawkers who escaped but returned to Death Valley 20 years later in 1869 to look for the Lost Gunsight and Silver Mountain. Though he found some silver, it was not of high enough quality to be the legendary gunsight. This rock with Rood's name and "1849" is believed to have been etched at the time of his original ordeal and still exists in Death Valley.

Resting Springs was a stop on the Old Spanish Trail where the forty-niners who stayed with Jefferson Hunt were refreshed by the plentiful water there. In the 1870s, a mine was established near the springs called the Gunsight, named for the famed Lost Gunsight silver. This 1890s photograph likely shows Philander Lee and his boys at the stone house at Resting Springs, which stands to this day.

After Addison Pratt completed a mission to Tahiti, to which he was assigned by Brigham Young, he returned to Salt Lake and was given a new mission. He joined Capt. Jefferson Hunt and the San Joaquin party and headed south to Los Angeles. When they left Resting Spring to make their way through the desert, Pratt made a discovery. (BYU.)

Near this site, named Salt Creek, Addison Pratt and others were looking for a good watering hole for camp. They noticed some loose quartz and began to search for gold. Soon, one of the men shouted, "Here is gold!" They took out a chisel and began to chip away some samples. The discovery caused excitement among the pioneers, as they had their first look at California gold. (MM.)

Uncle Chas Breyfogle Lost Mine.

This family photograph was captioned, "Uncle Charles Breyfogle Lost Mine." It is not clear whether the intent was descriptive or to convey frustration, as in, "Uncle Charles Breyfogle Lost [the doggoned] mine!" One thing is certain: the Lost Breyfogle gold became legendary throughout the West. In 1864, while running a hotel in central Nevada, Breyfogle met one of the original Death Valley pioneers who told him about some gold found near Death Valley. Breyfogle got a chunk of gold-bearing quartz and had a vague idea as to where it was found. He took a group to find the ledge, but he abandoned them and the rest of the party and headed back to Los Angeles. This pattern seemed to repeat with each group that went out with him. One time, he was discovered by a wagon train at Las Vegas Spring nearly dead. He claimed he was attacked by Indians for his gold. Years later, the Paiute claimed they saw him and were fascinated by his shoes, so they clubbed him and took them. They also carried him to the spring so he would survive. (JPB.)

RIVER REVEILLE.

ESDAY EVENING, MAY 29, 1866.   NO. 7.

LOCAL AND MINING.

NOTES ON TWIN RIVER.—We published
yesterday the fact of an important discovery having just been made in the North
Twin River District, by Messrs. E. S. Davis,
Russell Scott, and others. We are enabled
to give some additional details respecting
the newly-found mine, as well as of several others discovered and located by the
same parties. The Canada is situated in
the North Twin River District, near the

Tunnels are being run into the hillside, to
pierce both ledges forty feet below the
surface. Sufficient work will, undoubt-
edly, develop several of these large veins
into fine tangible property.

TURNED UP.—We learn from a letter
written at Pahranagat, May 13th, by W.
T. Nichols, that Breyfogle and party had
just arrived there from Los Angeles, via
Death Valley. It will be recollected that
for a couple of years this Breyfogle, whose
ways are as uncertain as a woman's, but
not nearly as pleasant, has been in search
of a fabulously rich gold mine, supposed
to exist somewhere in the vicinity of Death
Valley, the last discoverer of which is
now, unhappily for the Breyfogles, gathered unto his fathers. Somehow or other,
people who discover immensely rich
mines, scatter, and go abroad, and
are short-lived, and generally die without so much as a sign of the locality of
the treasure. But Breyfogle has large
faith, and is persistent, and means to find
it. He has already engaged in several expeditions in quest of the mine, but we believe he has never yet reached the neigh-
borhood of "good signs"—the scent has
always been cold. On one of these occa-
sions he lost his party, who had become
clamorous by repeated disappointment,
and wandered off alone, and was robbed
and half-scalped by Indians. He was
found in woful plight, by a party traveling to Salt Lake City, who gave him re-
lief, and carried him in with them. Since
which we have heard of him from every
point of the compass, and latterly at Pahranagat.

In late 1864, the editor of the Austin, Nevada, newspaper *Reese River Reveille* gave Charles Breyfogle $700 to find gold (this is called a grubstake). This 1866 article confirms that Breyfogle was still looking for the fabulous lost ledge and that he had taken many parties to search, but the end result was always the same. He seemed to routinely leave the parties he led into Death Valley, returning without any gold. (TF.)

Panamint Tom Wilson told the story of how two men searching for the Lost Breyfogle were lured to these hills near Daylight Springs called the Death Valley Buttes. Here, the men were killed by some renegade Paiutes from the Kern River area. Author Richard Lingenfelter states, "The Panamint Shoshone drove the intruding Paiute out and buried the two white men." Those who searched for this lost gold were called Breyfoglers.

In 1866, Nevada governor Henry G. Blasdel went looking for a shortcut from Silver Peak to the Pahranagat mines in Eastern Nevada. He ended up in Death Valley. There, he met Charles Breyfogle with his men at Furnace Creek. While resting there, Blasdel's men kept a keen eye on Breyfogle to see if he might give a hint as to the location of his fabulous gold ledge. No such luck—Breyfogle's gold remained a mystery. (NHS.)

Darwin French (standing, in suit) owned the Tejon Ranch and was the first American encountered by William Lewis Manly and John Rogers at the San Fernando Mission. In 1860, French led a party to find the Lost Gunsight mine. Along the way, he named places like Darwin Falls, Furnace Creek, the Panamint Mountains, and Towne Pass (named for one of the forty-niners, Captain Town). French's party never found the famed Silver Mountain. (MM.)

Dennis Searles accompanied Darwin French on his quest into Death Valley to find the Lost Gunsight. Though Dennis never found the Silver Mountain, four years later, he and his brother John found borax on what is now called Searles Lake, and by 1872, they had set up operations. Dennis is shown after stopping for lunch near what are known as the Trona Pinnacles. (CHS.)

Carl Glasscock, in his book *Here's Death Valley*, describes a scene from the Darwin French expedition: "Dry to the bone again they came to the abandoned camping place of the Bennetts and Arcans. Here were wagons falling to pieces in the desiccating sun, ox-yokes, chains, remnants of cloth, ashes of campfires, even crude toys abandoned by children. More important to the newcomers in July, there was water. They named it Bennett's Well."

The Darwin French party was disappointed. Yet they, like the 49ers, walked over and past white, salty playas like this that would later be mined for borax. Borax has hundreds of uses, including as a disinfectant, preservative, pesticide, and laundry additive. It became Death Valley's most valuable mineral, and unlike the elusive gold and silver ledges, it had always been in plain sight.

Motivated by the discovery of Nevada's Comstock Silver lode, prospectors scoured the Panamint Mountains looking for the Lost Gunsight silver. Though the lost mine itself was not found, it led to other happenings, including the 1873 discovery of a rich ledge of silver 10 miles up the aptly named Surprise Canyon in the Panamint range, and the founding of Panamint City, which is pictured here in 1875.

The Panamint City mining operation was overshadowed by a gang of thieves who had followed the original discoverers, which resulted in negotiating with the criminals and allowing them to stake claims. The silver ledge seemed to be rich enough for everyone, so one of the discoverers claimed the Wonder of the World, the other staked Stewart's Wonder, and the criminals staked out the Wyoming, which became the richest claim of them all. Panamint City is pictured above in 1875. Nevada senators John Paul Jones (below left) and William Stewart (below right) bought all of the claims in the summer of 1874 for more than $250,000 ($6.2 million in today's money). That started the rush and shouts of "Panamint! Panamint! Panamint!" Jones and Stewart did all they could to keep up the appearance of the world's richest silver strike. (Below, CHS.)

John Paul Jones and William Stewart sent ore to England to entice investors and constructed a mill and furnace built with more than 500,000 bricks. By 1875, Panamint City had grown from just a few hundred people to more than 2,000 and was the largest town within 100 miles. It was a tough town, where every other business was a saloon, and its homicide rate was among the highest in the West.

The Panamint City mill is pictured here in full production in 1875. When thieves threatened to steal silver from the freight wagons, William Stewart had the silver cast into ingots so big that no thief could lift one. The plan worked, and wagons carried silver like regular freight. But Stewart and John Paul Jones were previously involved in a mining scam, and when a new investigation was launched in 1876, it scared away Panamint investors. Stewart and Jones admitted the mines were played out, and Panamint was dead.

This 1907 article from the *San Francisco Call* summarizes the eternal quest for lost treasure: "The deserts have been assailed by forlorn hopes, by fortune hunters . . . who have found, only to lose again, the fabled Breyfogle, Pegleg and Lost Bethune." The article features quotes from Sen. William Stewart 30 years after the creation of Panamint City. There is a reference to the Lost Bethune, a reputedly lost ledge used to sell worthless mining stock. The Lost Pegleg is said to be in the deserts near San Diego. The stories are all pretty much the same. Someone finds gold or silver but cannot remember where, and they mount expeditions to find it and never do. Author Richard Lingenfelter writes, "The search was the thing . . . no one will ever know how many looked, and how many died. Risk was part of the lure. By 1893 it was said that the search for the Breyfogle had taken sixty lives, and in 1907 alone sixty men were said to have died. One journalist even reported that Death Valley had been named because of all the Breyfoglers who had died there! Yet some did find some real mines." (TF.)

Three

SINGLE BLANKET JACKASS PROSPECTORS

The desert prospector set out on the quest for riches with a pick, shovel, canteen, coffee pot, grub (food), a blanket, and a jackass, hence the nickname "single blanket jackass prospector." Unlike those who searched for riches in well-watered terrain (such as California's gold country), the desert prospector often faced life-threatening circumstances due to the desert's extraordinarily dry environment. The ability of the jackass to thrive in the desert as well as to sense the nearest watering hole became essential for the survival of the prospector.

If the prospector found a big strike, they did not hold on to it for long. They often went on a grand spree, which consisted of a trip to the nearest bar, drinking plenty of booze, and engaging with available women.

Most prospectors loved this way of life. They loved the quest and being out in nature, as it provided a sense of fulfillment. Attaining wealth was not their priority—it was simply to live the life of a prospector. They sold their claims and left mine development to others. They rarely invested their money or bought fine things. The prospector was addicted to the thrill of the find. In short, they had "gold fever."

Robert Service sums up this life in his poem *The Prospector*:

It was my dream that made it good, my dream that made me go
To lands of dread and death despised by man;
But oh, I've known a glory that their hearts will never know,
When I picked the first big nugget from my pan.
Yet look you, if I find the stuff it's just like so much dirt;
I fling it to the four winds like a child.
It's wine and painted women and the things that do me hurt,
Till I crawl back, beggared, broken, to the Wild.
For once you've panned the speckled sand and seen the bonny dust,
Its peerless brightness blinds you like a spell;
It's little else you care about; you go because you must,
And you feel that you could follow it to hell.

The following list of necessaries by Mr. A. Balch in his "Treatise on Mining" is as full as can be given by any one, and is more than the average prospector generally needs.

A PROSPECTOR AND HIS OUTFIT.

"*First*. Two pairs of heavy blankets weighing about 8 pounds each.

Second. A buffalo robe or a blanket lined poncho.

Third. Suit of strong gray woolen clothes, pair of brown jean trousers, a change of woolen underclothing, woolen socks, pair of heavy boots, soft felt hat, three or four large

Published in 1895, *Prospecting for Gold and Silver* states, "The regular prospector, at some time of his checkered career learned . . . how the veins appear on the surface, how to open a vein and the uses of pick, shovel and blasting powder. In a word he is a miner who has become too restless to stick to steady work and follows the uncertain and precarious livelihood of seeking for new and undiscovered veins." (TF.)

colored handkerchiefs, a pair of buckskin gauntlets, toilet articles, etc. All should go into a strong canvas bag.

Fourth. A breech loading rifle or shot gun and a revolver. Around his waist a strong sash to carry his holster and knife, in a sheath. His ammunition, if his revolver is large bore, may conveniently fit both his rifle and revolver. Pipe and tobacco.

Fifth. A sure footed native or mountain pony. A Mexican saddle with its saddle horn, straps, etc., to tie on various things, such as his pack, bags, water canteen, etc. The left stirrup may be fitted with a leather tube, in which the rifle barrel may be placed. A strap around the saddle horn will secure the gun stock. The long lariat or stake

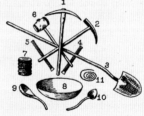

A PROSPECTOR'S TOOLS.

1, 2. Picks.	8. Pan.
3. Long handled Shovel.	9. Horn Spoon.
4, 5. Drills.	10. Iron Spoon.
6. Heavy Hammer.	11. Fuse.
7. Blasting Powder.	

rope for tethering his horse should be coiled up and tied by a strap to the saddle horn.

Sixth. For prospecting, a 'poll' pick and prospecting pan made of iron or a horn spoon should be carried. The pan is also useful besides for washing out sand, as a dish or bathing vessel. A large iron spoon for melting certain metals is likewise to be carried, and in some cases a small portable Battersea assaying furnace.

Seventh. A frying pan 8 inches diameter of wrought iron, a coffee pot, tin cup, spoon, and fork, and matches in tin box, pocket compass, a spy glass, or pair of field glasses.

The 1895 publication *Prospecting for Gold and Silver* continues, "These . . . ledges may make him in a moment a . . . rich man, and if he finds them, they will cost only a simple compliance with the inexpensive regulations of the law. So the life of a prospector offers many attractions to one who is restless and loves to roam and loves to find something new and is not afraid of considerable hardship." (TF.)

In this glass-plate image from the 1890s, this grizzled prospector on his trusted burro is likely reviewing a claim notice with the two men at left. After making a strike, it was important for a prospector to put markers in the ground to stake his claim and then head to the county recorder to have it registered. Death Valley's county recorder was in Independence, more than 100 miles west.

This 1902 image shows the south end of Death Valley, where the *San Francisco Chronicle* claimed that "there is enough niter in Death Valley to make it a center of population and life instead of the center of desolation and death." However, one geologist responded, "It strikes me as a little short of insanity for the average miner to go into Death Valley to locate niter claims." (CSL.)

Niter, also known as saltpeter, was used in both gunpowder and fertilizer. In 1910, San Francisco businessman Albert Scott acquired over 13 square miles around Saratoga Springs in Death Valley's south end. Scott's prospecting settlement at Saratoga Springs included not only tents but a wooden structure. In this image, it seems that one of the men has just arrived with his bedding

and canteen. The lady of the camp is looking on from the shade of the shed door. Based on her demeanor and attire, she looks to be in charge. Note the burro in the background at left. There is a washbasin near the tent, and the large drum likely contains water. Most of the men's shoes are white due to the salt- or alkali-filled terrain of the area.

The men pictured were some of the prospectors hired by Albert Scott to find and develop niter deposits. Scott sold more than $200,000 in stock and built some cabins at Saratoga Springs. Scott's development caused a rush, sending more than 1,000 hopefuls into the south end of the valley to look for niter. Needless to say, that petered out.

These men, with a buckboard in the background at left, are part of the California Nitrate Development Company formed by Albert Scott. Outfitted as typical prospectors, the men seem to be in a rather good mood as they stand in front of their tents near Saratoga Springs. Note the rock foundation for the tents, indicating that this is not just an overnight camp.

Loading up to go out into the backcountry was never easy. Here, a prospector loads everything he will need on the back of his mule. While burros and donkeys also accompanied prospectors, this is a mule. A mule is a cross between a male donkey and a female horse. Although mules were often used for packing and pulling wagons, most prospectors preferred the smaller and more compliant burros.

Once his mule was loaded up, the prospector was ready to head into the desert. In the rocky and dry terrain of Death Valley, it was important to know where one could get water. Prospectors often learned from Native Americans where to find water and the most likely places for springs.

George Cook (in the middle of the center tent), like many in the desert, had a job and also prospected. Cook was a freighter who often transported wood for the mining camps. Here, he is pictured in 1906 just east of Beatty, Nevada, at a camp called Transvaal while on a prospecting trip.

From the late 1800s through the 1920s, scenes like this were common in Death Valley country. Carl Glasscock writes that the burro is "a noble beast in heart, stamina and adaptability. Like the prospector who commands him, the burro may lie and steal, but he persists. He gets there regardless of all obstacles, as gaunt and hungry and thirsty as his master, but still on his feet."

Here, a prospector demonstrates the use of the pick. The pick and shovel were as important as the burro for the desert prospector. Once the prospector struck some rich ore, he would make a claim marker. He might use stones or poles to outline the extent of his claim. Then, he would travel to the county recorder to document his claim.

Once a prospector returned with his recorded claim paperwork, he would usually put it inside of a tin cigar can—often a Prince Albert can. He would take the can and put it in the pile of stones, and the claim would legally be his. The stone marker and the paperwork contained within it served as a notice to anyone who found ore in the area that a claim had been made on the land.

Here, a prospector has brought some of his rock to have it assayed or evaluated for how rich it is. If it is determined that the claim has good ore in it, then the prospector may sell the claim. If it is purchased by a promoter, then the promoter tries to sell it to a mining company or an investor

Once a property was sold or investors were brought in, then mining could begin. This photograph shows the early stages of mine development in an area where it appears that there is good ore on

who will develop the mine. Death Valley's remoteness also provided an opportunity for promoters to sell high-priced stock in worthless mines.

the claim. The miners are in the beginning phase of determining value and the extent or length of the vein of gold or silver.

Frank "Shorty" Harris (left) was born in Rhode Island on July 21, 1857, and headed west as a teenager in the 1870s to seek his fortune in the mines. Harris came to embody the essence of Death Valley's single blanket jackass prospector. He found and lost several fortunes and had a weakness for the "Oh Be Joyful"—a slang term for liquor. Pictured here in what is called the Devil's Golf Course in the heart of Death Valley, Harris lived in Panamint Valley to the west. From his home in the town of Ballarat, he would go out and prospect in the desert. (ECM.)

Shorty Harris's most famous find happened on August 9, 1904, just northeast of Death Valley proper, with Eddie Cross (right). According to historian Richard Lingenfelter, "They never could agree on who picked up the first piece. Ed said he spotted that glistening rock, about the size and color of a bullfrog, and called to Shorty who only smiled skeptically. But as Shorty began to examine it his cheeks flushed with excitement and he finally let out a war whoop, jumped up and shouted: 'Hellfire, Eddie we've struck the richest jackpot this side of the Klondike!'" This was the beginning of the Bullfrog rush. (NHS.)

Here, Shorty Harris is on the dunes near Stovepipe Wells. In 1904, when Harris and Eddie Cross struck gold in a green rock that looked like a bullfrog, they named their claim Bullfrog. In Goldfield, they learned their ore was so rich in gold that when word got out, it started a rush! Harris headed to the bar, got drunk, and sold his interest for $1,000. When he sobered up, he blew it all by treating friends to a few more rounds. (ECM.)

Shorty Harris loved his burros and would brag about how smart they were. It is a fact that burros can smell water and have saved more than one prospector by taking them to a spring. Of course, Harris also had his tall tales, such as the one about a burro that had a toothache, disappeared one night, and came back with a gold filling. Here, Harris waters his burros. (ECM.)

The other big strike Shorty Harris was partly responsible for was made on the western side of Death Valley in the Panamints. Harris and Jean Pierre "Pete" Aguereberry were prospecting in July 1905 when Aguereberry stopped to look at a ledge that prospectors had passed for decades. He chipped off a piece and found it was rich in gold. Harris, pictured here around 1930 in his later years, was skeptical but saw that the rock was rich indeed.

Once they got to Ballarat, Shorty Harris went to a bar—possibly the Black Metal Saloon. While drinking the Oh Be Joyful, he began to brag about the rich strike he and Pete Aguereberry made and revealed exactly where it was. By the time they got back up to their claims, there were about 300 men they had to keep from jumping their riches. A claim jumper is someone who steals another prospector's claims. (ECM.)

Shorty Harris sold his share in the 1905 claims for $10,000, the biggest single payday of his life. Pete Aguereberry (pictured) got some stock for his share but not much cash. Aguereberry and Harris had agreed they would share the name of the new camp, calling it Harrisbury to honor both of them, but it was soon changed to Harrisburg, leaving out Aguereberry entirely.

The camp at Harrisburg (which should have been Harrisbury) never really amounted to much. There was one tent saloon and a store. Water was hauled in from Emigrant Spring, which was about eight miles away. By the winter of 1906, the excitement was gone, although some hangers-on remained, as shown in this 1908 image.

Pete Aguereberry may not have struck his fortune in gold, but he gave visitors a long-lasting gift when he discovered a magnificent vista of Death Valley and built a road leading to it. Aguereberry lived near this view for more than 40 years. His cabin survives to this day within Death Valley National Park, where people can still visit Aguereberry Point.

In some ways, the prospector was as much a mascot for Death Valley as the cowboy was for Dodge City and other cow towns of the West. Although it was perhaps not as romantic as the ideal of the cowboy and his horse, the grizzled prospector who roamed Death Valley with his burros could be just as rough and tumble—and drink and gamble as hard—as any Texas cowboy.

Alice "Happy Days" Diminy headed for the Death Valley gold rush in 1906. She ran restaurants and a saloon called Happy Days, which was so named after a satisfied miner who ate a meal at her restaurant declared, "Oh, happy day!" Diminy described the life of the prospector this way: "Did you ever suddenly find yourself a partner in a gold mine? There's a thrill. You don't know what lays ahead, but you see yourself rolling in wealth."

If a prospector wanted to develop their claim, they might use a drill to make a hole for explosives. This old prospector has a drill in one hand and the hammer in the other. Drilling was achieved by alternately hitting the drill and turning it. One man using this tool could drill an average of 6 to 10 feet in a 10-hour day, with possible drilling rates of up to three inches per minute. (TF.)

One man with a drill was called a single jack. If he had help, where one man (called a shaker) held the drill and the other hammered, it was called a double jack. Hard-rock mining could be both tedious and dangerous, particularly when it involved the use of explosives.

Washing Down The Alkali

Death Valley's salty desert floor is often referred to as alkali, as noted in the caption. This prospector's rig is fairly elaborate, with two burros pulling the wagon (and a dog) and his prospecting burro in tow. Water could become more precious than gold, as new prospectors, called greenhorns or tenderfoots, were often unprepared.

Shorty Harris (right) prospected from the days of the Wild West until the era of the automobile. Here, he sits for a meal with a fellow prospector with the flivver (car) and Death Valley in the background. When the automobile appeared in the desert, prospectors argued whether it was best to use burros or the newfangled flivvers. Most stuck with burros, as no flivver could go where the trusted burro was able to venture. (USSC.)

In this photograph taken 16 months after the discovery of the Bullfrog lode by Shorty Harris and Ed Cross, wood-frame buildings are visible in the new mining town as it was growing to become a hub for prospectors to explore Death Valley country. While the identities of the prospectors shown here are unknown, this is a rare iconic image that features both an experienced prospector and some greenhorns. The middle burro carries the tent, and the outfit appears well supplied for a journey of several days. Author Bourke Lee, in his book *Death Valley*, says, "Prospectors were crawling through the naked country, looking for treasure. And they were finding it. They found rich silver and gold ore. They built cabins in the brief watered areas of the Death Valley region. They did a little work on their claims and those who had high-grade [rich ore], packed enough out to promote grubstakes and keep themselves in bacon and beans. They cursed the country but they stayed. They liked it." (TEC.)

Four

DEADLY QUEST

The quest to find gold in Death Valley attracted seasoned desert prospectors as well as new adventurers. Many of the greenhorns, from temperate climates with an abundance of water, were unaware of the desert's dangers. They underestimated distances, had no knowledge of water sources, and ventured out to seek their fortunes only to return in a box or be buried where they lay.

According to Arthur J. Burdick, author of *The Mystic Mid-Region*, "The remarkable mineral wealth of the region has been a glittering bait to lure men to destruction. There are in the valley golden ledges, the ores of which run in value to fabulous sums per ton. There are vast beds of borax, niter, soda, salt and other mineral drugs. The Valley is a storehouse of wealth, the treasure-vault of the nation, the drug-store of the universe, but Death holds the title."

Desert men succumbed as well, becoming too confident in their abilities and ignoring common precautions they had always urged others to take. There were other causes of death, such as illness, tainted food, accidents, suicide, homicide, and one sensational tale of a lynching in a mining camp. However, fully two-thirds of all of Death Valley's fatalities are the result of its brutal environment. Death in the valley does not happen due to it being extremely hot but because it is extremely dry.

"On an average summer day in Death Valley, you can lose over 2 gallons of water just sitting in the shade; hiking in the sun you can lose twice as much," historian Richard Lingenfelter writes. "Without enough to drink to replace it, the loss of 4 gallons of water is almost certainly fatal, and even the loss of 2 gallons could have fatal results!"

It is a miserable way to go. The tongue swells, balance is lost, and delirium sets in. The sick individual will take off their clothes and mindlessly dig for water. Their tongue and skin shrivel, their vision dims, their hearing goes, bloody cracks appear in their skin, and soon, death arrives. The region's deadliest year was 1905, when more than 30 souls perished during the height of the Death Valley gold rush.

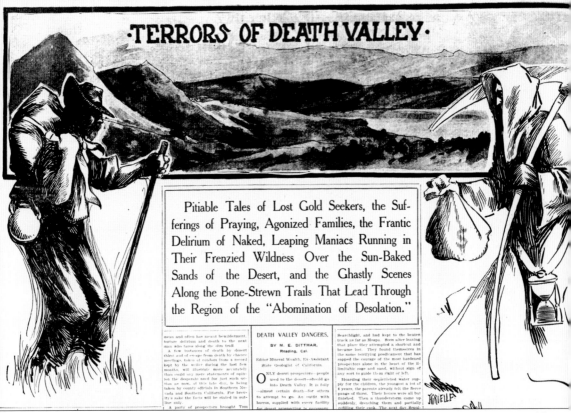

·TERRORS OF DEATH VALLEY·

Pitiable Tales of Lost Gold Seekers, the Suf-
ferings of Praying, Agonized Families, the Frantic
Delirium of Naked, Leaping Maniacs Running in
Their Frenzied Wildness Over the Sun-Baked
Sands of the Desert, and the Ghastly Scenes
Along the Bone-Strewn Trails That Lead Through
the Region of the "Abomination of Desolation."

DEATH VALLEY DANGERS.

BY M. E. DITTMAR,
Reading, Cal.

Editor Mineral Wealth, Ex-Assistant
State Geologist of California.

ONLY desert prospectors—people
used to the desert—should go
into Death Valley. It is folly
—almost certain death—for others
to attempt to go. An outfit with
burros, supplied with every facility
for desert prospecting is necessary.

In places far from Death Valley, such as Muncie, Indiana, its horrific stories and tales were told to a fascinated public. Its reputation as the "pit of horrors," as one New York paper described it, or the "Region of the Abomination of Desolation," as this *Muncie Star Press* article states, thrilled readers across the country. This article appeared in 1905 at the height of Death Valley's fame and notoriety. The images of bleached bones in the desert, naked thirst-crazed maniacs running through the sands, and other scenes that can only be described as ghastly were common due to the name the forty-niners had placed on the land. But it was true that Death Valley could be deadly. Although the article may be overstating when it says, "Only desert prospectors . . . should go into Death Valley. It is folly—almost certain death—for others to attempt to go," there is no doubt that those who were ill-prepared or took the desert for granted could meet a sad end. The writer of the article cleverly finishes by stating, "It is Death Valley. And beside it lies Funeral Range."

This 1904 photograph of a man who perished in Death Valley must have been taken soon after his death, as neither the elements nor the animals have yet taken their toll. The *Muncie Star Press* reported in 1905 that "once off the trails and out of water, the desert traveler goes mad, wanders aimlessly among the rocks and sand hills and lies down to die in some hollow where his whitened bones may rest for years undiscovered."

The elements and the animals have ravaged the unidentified body in this 1907 picture that included an apt descriptive note on the back—"grisly corpse." Prospectors going out alone often got disoriented, as many of the canyons and mountains of Death Valley look similar. Misjudged distances—along with an inadequate water supply without knowledge of the locations of springs— can have dire results.

GRAVE OF MAN WHO PERISHED IN DEATH VALLEY AND BONES OF HIS 6 HORSES

Perhaps the most famous grave in Death Valley is that of Jimmy Dayton. He came to manage the Greenland Ranch (now Furnace Creek Ranch), growing alfalfa for William Tell Coleman's borax operation during the days when the big teams hauled from Harmony Borax Works to the railroad at Mojave, California. In July 1899, Dayton only made it 20 miles south of Furnace Creek, where he and his six mules died.

According to Carl Glasscock in *Here's Death Valley*, When Jimmy Dayton's wife moved to Los Angeles, he resigned from managing the ranch and set out with a wagon, six mules, and his dog to join her. When Dayton did not arrive in Daggett, some 150 miles away, the teamster Frank Tilton went looking for him along with longtime Death Valley resident Adolph Nevares. They found his body "twisted a bit as if in pain," wrote Glasscock. "Jimmie's dog, starved to a skeleton, unharmed except for hunger." Here, an explorer pauses at Dayton's grave. (CSL.)

"BURY ME BESIDE JIM DAYTON IN THE VALLEY WE LOVED. ABOVE ME WRITE: 'HERE LIES SHORTY HARRIS, A SINGLE BLANKET JACKASS PROSPECTOR."- EPITAPH REQUESTED BY SHORTY (FRANK) HARRIS BELOVED GOLD HUNTER. 1856-1934. HERE JAS. DAYTON, PIONEER, PERISHED, 1898

TO THESE TRAILMAKERS WHOSE COURAGE MATCHED THE DANGERS OF THE LAND, THIS BIT OF EARTH IS DEDICATED. FOREVER.

DAYTON

HARRIS

THE LAST RESTING PLACE OF JAMES DAYTON AND SHORTY HARRIS, DEATH VALLEY NAT'L MONUMENT, CALIF.

When prospector Shorty Harris died in 1934, he wanted to be buried next to his best friend Jimmy Dayton. The original wooden headboard was actually an ironing board that belonged to Pauline Gower, wife of borax mine superintendent Harry Gower. State senator Charley Brown paid for the plaque on the stone monument, though the date of Dayton's death was incorrect. According to those who attended Harris's funeral, the grave was deep enough but not long enough for the casket, so the attendees had to wait as the hole was extended. (FF.)

RIFF BEYOND RELIEF

James Riff, a carpenter from Bullfrog, is the last reported victim of Death valley.

Andy Johnson and James Riff, with three burros, left here last Saturday on the perilous trip in Death valley. Two or three rescuing parties went out, but at the hour of going to press Riff had not been found.

Riff and Johnson stopped at Keane Springs Saturday night and Sunday morning started on a prospecting trip to Grapevine via Hole-in-the-Rock. They reached Tule springs on Monday. Here they filled their canteens and started back to Hole-in-the-Rock by a different route from that which they came. The water ran out at 5 o'clock in the afternoon.

Johnson and Riff were together until 9 o'clock Tuesday morning. They separated at a point about equi-distant from Hole-in-the-Rock, Tule springs, Stovepipe springs and Triangle springs, where Riff faltered and fell. He was unable to go further, but urged Johnson to rush on for help, saying he would pay the necessary expense. Johnson made Keane springs Tuesday, traveling by way of Furnace creek and Hole-in-the-Rock.

Immediately upon hearing Johnson's story the miners of Keane springs organized a rescuing party and started out. Messrs. Dart, Ryan, Derat and Matteson constituted the party. Another man with a bloodhound later joined the rescuers. The party started out that (Tuesday) night.

Wednesday morning C. Kyle Smith, the recorder, loaded Johnson in a wagon and took him to where he thought he had left Riff. They saw where there had been a man laying down under a bank one-fourth of a mile from where Johnson supposed he had left his companion. Tracks from this point were followed down as far as the Death valley Buttes, near Stovepipe springs, the tracks appearing to lead towards this last named water hole.

Mr. Smith said that as soon as the sun went down and it became cooler the man with the dog was going to put the hound on the trail.

Hole-in-the-Rock is headquarters for the rescuing party.

Yesterday Matteson and a friend from Bullfrog again took up the search. Riff is spoken of very highly.

E. C. Dart of the rescuing party, came to Bullfrog this (Friday) morning. Mr. Dart reports that the rescuers have all returned to Keane springs without having located Riff, who, in Mr. Dart's judgment, wandered from the spot where Johnson left him to the small washes towards Surveyor's well, where it is exceedingly difficult to find the body.

SURVEY STARTED

Altitude In Front of the Bullfrog Miner Office 3,573 Feet

Work on the governmental geological survey of the Bullfrog district was started yesterday morning. Deputy Ross Phillips, in charge, estimates that the district will be covered in two weeks.

Topographer R. H. Chapman is in charge of the Nevada work as a whole. His territory embraces about 9,000 square miles and, roughly stated, takes in the country from Tonopah to Las Vegas. The work is being done by latitude and longitude.

The Bullfrog special map, as well as that for Goldfield, will be drawn on a large scale. The altitude of Bullfrog, taken in front of The Miner office, is 3,573 feet above sea level.

Deep in Death Valley National Park's archives are letters, newspaper articles, and photographs related to the 1905 death of James Riff, "a victim of Death Valley." Writing to Riff's family, Riff's business partner Paul Wright stated, "Andy Johnson owned an outfit and asked Jim to go to Death Valley with him prospecting. When their canteens became empty, Riff was exhausted and unable to travel." Johnson left Riff to get water and food, but when he returned, Riff was dead.

Paul Wright explained about his friend James Riff: "It was . . . perhaps 120 or thereabouts in the shade. But when his mind left him, as it does all who perish in the trackless sand, he got up and walked straight towards the melting sand of Death Valley. I expect to take a team and driver to haul water and a friend and a coffin and go to Death Valley." Pictured here are the hearse and water wagons assembled by Wright.

A prospector from Colorado found James Riff's body. "I took an undertaker with me," wrote Paul Wright. "He fixed the body very nicely. I will send you pictures. The undertaker bill was $50, the labor bill was $5 a day, two days. The man looking for his body was $83. If [Riff's] people wish to pay a portion of this, of course I would except [*sic*] as perhaps they are more able than I am."

"I will send you pictures of the grave when they are finished," Paul Wright wrote to James Riff's father. "Jim's earthly possessions here in Rhyolite will probably amount to $50 . . . but he often said, 'I will make a stake or die trying,' and he has made his word good, as he did in everything else he promised to do."

The caption with this photograph states that it is a "view of skeletons of Tommy Flaus and Emory Bodge who died of thirst on the desert in Death Valley while attempting to cross it on foot from Rhyolite to Skidoo in August 1909. They were found about three weeks after their death by a search party from Skidoo."

Dr. Reginald MacDonald is shown holding the entangled skeletons of Emory Bodge and Tommy Flaus. MacDonald was an itinerant doctor who served mining camps from Randsburg (three hours northeast of Los Angeles) to Ballarat and Death Valley. He came to the desert looking for the Lost Pegleg Mine. His life motto was shown on a plaque on his wall: "I live each day so I can look any man in the eye and tell him to go to hell."

In 1908, the mining camp of Skidoo, located in the mountains above Death Valley, was going strong. In April of that year, it was here that one of Death Valley's most notorious events took place. It involved two deaths and the macabre ongoing story of a man who murdered one of the town's most respected citizens.

Joe "Hootch" Simpson is shown standing in the back of his restaurant in Gold Reef, Nevada, before coming to Skidoo. In Skidoo, he ran the Gold Seal saloon and was reputed to be a pimp for "Skidoo Babe." Simpson was being treated by Dr. Reginald MacDonald for syphilis, and at one time, MacDonald patched up Simpson's nose after he got into a shoot-out and had part of it shot off. Simpson was notoriously a mean drunk. (NHS.)

On Easter Sunday morning, April 19, 1908, Hootch Simpson was drunk and went into the Skidoo Trading Company store, where its owner, Jimmy Arnold (right), was behind the counter. Simpson pointed a gun at Arnold and asked, "What have you against me?" Arnold replied, "Nothing, Joe." Hootch replied, "Prepare to die. I'm going to kill you," and started firing.

On that fateful Easter Sunday in 1908 when Hootch Simpson shot Jimmy Arnold, Dr. Reginald MacDonald (center) was across the street and heard the shouting but could not get there in time to save Arnold. MacDonald drew his gun and told Simpson to drop his and surrender. But Simpson took a shot at the constable and threatened others who were gathering. Finally, they disarmed him and took him to a shed to await the sheriff, who was coming from Independence.

SKIDOO NEWS.

VOL. II, NO 18   SKIDOO, INYO COUNTY, CALIFORNIA, SATURDAY, APRIL 25, 1908   TEN CENTS

MURDER IN CAMP
Murderer Lynched
WITH GENERAL APPROVAL

Joe Simpson Shoots Jim Arno'd Dead and Is Hanged By Citizens

The disturbance which has shaken this community to the roots, in the past few days, opened on Sunday morning last, at about eleven o'clock, when Joe Simpson, familiarly known as Joe "Hootch" (that being his favor to beverage) held up the Southern Calif. Bank here, for the nimble sum of twenty dollars, that being the sum of his immediate need. He was overpowered before he could collect, and his gun taken from him. He returned to the bank (which is located in the store) again and became very abusive Jim Arnold, managing partner in the store, finally put him out. Three hours later, he returned again with his gun and deliberately shot Arnold, who was unarmed. He turned, and covered the banker, Ralph E. Dobbs and would probably have killed him had not his identity been divulged. He was overpowered and handcuffed. Arnold died the same evening. An inquest on Arnold's body was held on Monday, the jury returning a verdict, "killed by gun-shot wound, inflicted by Joe Simpson." Sometime on Wednesday night an armed body of citizens overpowered the sheriff and seized prisoner and hanged him to a telephone pole. On Thursday, inquest was held on Simpson's body, the jury finding that "he died by

raised his gun and shot Arnold just below the heart. Turning quickly, he threw his gun on Mr. Dobbs behind the bank counter an' commanded him to come out an' die, but before he could fulfil his threat, he realized his own danger and backed out into the street. Simpson's entrance to the store and the "crack" of the shot, caused a scene, that, if a moment, was more dramatically tense, than ever pictured in play or story. Constable Sellers, who was reading a newspaper in the Club saloon, snatched up a shot gun and loaded as he ran, the shells wedged!! flinging it down, he leaned over and grabbed a six shooter from beneath the bar. The rapidity of his actions may be judged from the fact that he was up with the murderer, before he had crossed the street. Others were equally quick. At the sound of the shot Dr. Macdonald dashed into Pfluger's saloon and snatched up a long barrelled winchester rifle. It was a great

Sellers should be highly commended for the great bravery he displayed through out the action. He virtually carried his life in his hand, from the time he appeared upon the street, until he had subdued the murderer. He showed great patience too, with McBain. Many officers would have shot him down without argument.

Simpson handcuffed, but jubilant at his cowardly crime and at the hot fight he had put up, was taken to the Club saloon, until a guard-house could be decided upon.

DEATH OF ARNOLD

When Simpson's first shot tore its way through Arnold's vitals, he sank to the floor crying, "For God's sake don't shoot again Joe. You've got me now!" and in the excitement of the dramatic events of Simpson's capture, he was, for a moment, half-forgotten. Like a dying wild animal that bites itself, music crowded in and ran into the cellar below, his life ebbing away from internal hemorrhage. He was found and carried down to Dr. Macdonald's surgery. A cursory examination convinced the doctor that the chances were against the patient. The bullet had entered just below the heart and passing beneath the ninth rib, punctured the corner of the liver, sev-

tions about his interference, that so many resulted in other deaths. After deliberation, the jury, consisting of W. E. Follansbee, M. Gavrelaud, A. H. Swinerton, J. H. Wilson, C. J. Shackett, A. T. Hall, J. J. Sheeby, F. Pfluger and W. McCoy, brought in a verdict "that the deceased, James Arnold had died from the effects of a gun-shot wound, inflicted by Joseph L. Simpson."

Early in the day, the District Attorney was telegraphed for to take charge of the case. The funeral was arranged for the following day and it was then the widespread feeling of regret and deep respect in which the deceased was held, manifested itself.

A DESERT FUNERAL

At noon on Tuesday the victim was buried, with a simplicity and pathos, unknown in large communities. The casket, fashioned with loving hands in the carpenter's shop of the Skidoo store, was placed on a wagon suitable draped, and drawn out to the cemetery. In the absence of a cleric, A. T. Hall conducted the service, which was opened by the singing of "Rock of Ages" by a male quartette, composed of Charles J. Shackett, Thorleif Olson, S. W. Kline and Judge Frank Thiess. Following this, Mr. Hall read the Burial Service of the Church of Eng-

follows. "In the early part of the week, I feared violence, but as the days went off, I felt that the ill feeling had cooled off. On Wednesday evening, shortly before midnight, the door was broken in and Deputy Heath and myself were overpowered. A dozen guns were presented against us and we were told that if we made a sound, we would have to take the consequence. The guard-house, was a one-roomed building of this corrugated iron, in poor repair. The walls could be kicked in anywhere. Resistance was useless. The night was so dark I was unable to recognize one man from another. There was a large number present, about fifty I judge. Only two of the jury spoke. I could not recognize either voice. The prisoner was awakened taken fearless. Guards remained at the door and window to prevent us from coming out. We saw nothing of the lynching."

The body was discovered early next day, hanging and Judge Thiess adjusted of the fact. He immediately ordered the body to be cut down. An inquest was held later in the day, but no information could be obtained as to who the perpetrators were, the difficulty. While there was a general feeling of levity outside the court, investigation was conducted with due dignity. The judge evidently felt his position very keenly. Outside the court, several references were made

---

The Easter 1908 shooting of Jimmy Arnold by Joe "Hootch" Simpson was big news in Skidoo and Death Valley country and even made the *Los Angeles Times*. As the "Skidoovians" buried Arnold on the Tuesday after Easter, they sang "Rock of Ages" and "Nearer My God to Thee." But it was revenge that was on their minds, as Simpson was seen gloating and putting a notch in his gun after the shooting. Just around midnight on Wednesday evening, about 50 masked townsfolk took Simpson out and hanged him from a telephone pole. The next morning, Dr. Reginald MacDonald took the body to his tent and captured images of the corpse. He took one picture of the body on the table and another, shown at right, after he strung Hootch's body up in his tent. (Above, ECM.)

"Hooch" Simpson
Skidoo, Calif.
Easter Sunday
1909

Some mementos of the event still survive, such as the noose used to hang Hootch Simpson. On display at the Goffs Cultural Center in Goffs, California, which is run by the Mojave Desert Heritage and Cultural Association, the noose is a reminder of how fascinated people were with the event at the time. It is a relic of one of the last recorded lynchings in California. (MDHCA.)

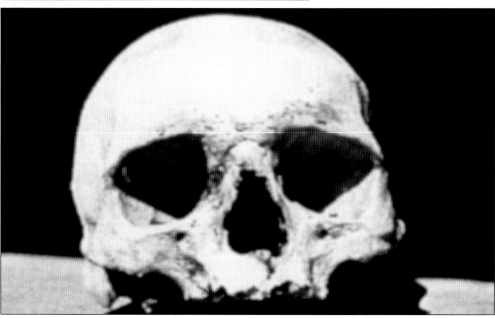

Dr. Reginald MacDonald was curious about the effects of syphilis on Hootch Simpson. After Simpson's body was tossed down a mine shaft, MacDonald retrieved it, cutting off the head. He boiled it and let ants eat the rest of the skin. He hid the skull in his tent floorboards, and in 1917, some prospectors found it and took it to a dentist in Trona, California. The skull changed hands several times until it came to rest with a collector near Joshua Tree, California. (TF.)

Five

EUREKA!

Every prospector who has put his pick or shovel in the ground and noticed a yellow glint shining in the rock has probably shouted the state motto of California—"Eureka!" Stemming from the Greek word, it means, "I have found it," and was likely first uttered by the mathematician Archimedes more than 2,000 years ago when he discovered how to determine the gold content of the king's crown. This was a crude form of the process every prospector and miner knows as assaying. If one wants to know how much gold is in a rock, it is the assayer's job to find out.

Of course, it does not have to be gold that evokes a "eureka" from a prospector—it could also be silver or copper. Death Valley's gold rush actually began with silver, and it was not technically located in Death Valley.

In 1900, about 130 miles north of Death Valley's Furnace Creek, a prospector named Jim Butler picked up a rock to get his burros moving and discovered that it was rich in silver. This started a rush, and the town of Tonopah, Nevada, was born. It was a lifesaver for Nevada, as there had not been a big strike since the Comstock Lode boom of 1859, and the state was in bad financial shape. As Tonopah grew, prospectors ventured into the Nevada desert. In 1902, with the help of a Shoshone man named Tom Fisherman, prospectors William Marsh and Harry Stimler found gold about 20 miles south of Tonopah. That discovery grew into Goldfield, which soon became Nevada's largest town.

Prospectors headed south, coming ever closer to Death Valley. In 1904, Shorty Harris and Eddie Cross made the discovery of a lifetime on the edge of Death Valley. It was not long before the dreaded valley was luring treasure-seekers from around the world.

It was possible, it seemed, for anyone to come to Death Valley and get rich. Promoters touted it as the "treasure basin of the world," where riches were there for the taking and one could dig and soon be shouting, "Eureka! I have found it!"

In her book *A Mine of Her Own*, Sally Zanjani describes Isabella "Belle" McCormick Butler, pictured here around 1900, as one of the "most successful women prospectors of all time." While Belle's husband, Jim Butler, tells the story of the rock and his burros, it was likely the Native Americans who told Jim where to look around the springs called Tonopah. Referred to at times as the "Lucky Lazy Rancher," Jim seemed in no hurry to explore his claims. It was Belle who decided that the claims "could be more satisfactorily located under her direction and finally got her husband to go explore the rich strike he had made." Belle was 40, and Jim was 45, when they headed to the springs at Tonopah. While looking for water and chasing burros, Belle prospected the area for about a week and then struck and claimed the Mizpah, which became the richest of all their claims and was the central mine responsible for helping to revive Nevada. When Belle died in 1922, she was eulogized as the "Mother of Nevada." (JBF.)

Belle (standing at left on top of ore sacks from the Mizpah Mine) and Jim Butler (seated next to Belle) are pictured in 1901 with their dog. Mizpah, named for the biblical term for a blessing, was Tonopah's richest mine. From 1900 to 1921, Tonopah's mines produced nearly $121 million in precious minerals, which is the equivalent of more than $3.5 billion today. The Butlers leased their claims to miners who worked in exchange for a share of the money; when big businesses took over the mines, the leasing ceased. (JBF.)

By 1902, Tonopah was a booming town. A young boy is peddling newspapers, and ore wagons are ready to hit the road. People were coming from all over the country to join the rush. The Klondike Rush in Alaska had played out, and those chasing the next big rush, like lawman Wyatt Earp and writer Jack London, headed to Tonopah. (JBF.)

There were African American communities in nearly all of the mining camps. This unidentified couple made their adobe home in Tonopah in 1902. Rocks, tin cans, and bottles appear to be holding up the house and lining the front yard. John Howell was a Black rancher near Beatty with one of the largest and most significant ranches in the region. When the railroads came to town, African Americans often arrived while working as porters. (CNHS.)

This photograph, which was turned into a postcard, captures the optimism and excitement that surrounded Tonopah. Prospectors used the town as a base from which to prospect the surrounding desert. Prospectors often got to know local Native Americans, and it was often those Native Americans who showed them to the location of the rich rock. When gold was discovered 20 miles south of Tonopah, it started the rush to Goldfield in 1902. (TF.)

Within a matter of four years, Goldfield became the biggest city in Nevada, with a population of more than 20,000. It was touted and known across the country as the next big American gold rush. From Goldfield, prospectors ventured further south, and when Shorty Harris and Ed Cross discovered rich gold in a rock that looked like a bullfrog in 1904, the rush to Bullfrog began. (TEC.)

Shorty Harris described the 1904 rush: "I've seen some gold rushes in my time that were hummers, but nothing like that stampede. Men were leaving town in a steady stream with buckboards, buggies, wagons and burros. It looked like the whole population of Goldfield was trying to move at once . . . there wasn't a horse or a wagon in town, outside of a few owned by the big companies!" Here, a stagecoach heads out of Goldfield.

In this 1905 image, a group of prospectors are shown assaying some rock they brought in. The burros in the background are feeding without their packs on. The open tent reveals a shelf with dishes, cups, and a coffeepot. Sticks have been gathered for a fire, and a canteen is on the ground. This and the following two photographs are some of the earliest known images of the Bullfrog mining district. (TEC.)

Pictured here are a woman named Jack and her dog Dutch. Most men in the mining camps were single, but Jack came with her husband, and their dog stayed with them in this crude cabin. On the right side of the image, it appears that there is a spigot attached to a can for water, which would have been a mining camp luxury, as water could be expensive in dry camps. (TEC.)

The husband of Jack (the woman pictured on the bottom of the previous page) is shown with his dog Dutch. Based on the way he and Jack are dressed, their wood cabin and running water, and the fact that he had a camera, it could be assumed that he had some money or was grubstaked by someone who did. (TEC.)

In the early days of the Bullfrog district, moving was as simple as picking up the tent to relocate it. Here, early settlers move a tent down the street. Note the wood-frame platform that made it possible to lift the canvas tent. The structures were often hybrids of wood-frame floors and partial walls with canvas on the sides and ceiling. They did get hot in summer. (CNHS.)

This 1905 image shows a stage arriving in Bullfrog. The tent building behind the stage is called the Northern, which seemed to be a common name for saloons and hotels at the time. To the left

is a more substantial wood building advertising itself as "Restaurant." The telegraph pole shows that even the most remote camps tried to stay in communication with civilization.

The *St. Louis Post Dispatch* of December 10, 1905, carried this colorful image of Christmas Eve in Bullfrog. The letters "ARO" in the background indicate the faro table; faro was a popular gambling game at the time. There is no record of a Green Head Saloon in Bullfrog, and the photograph is likely staged, providing further evidence of the fascination and curiosity this area inspired as the "last holdout" of the Wild West. (TF.)

The last great gold rush was unique because there were automobile stages, such as this seven-passenger Pope Toledo. It took passengers from Bullfrog to Goldfield in just five hours for $25—a rather expensive ride at nearly $800 in today's money. The stage's most famous driver was William Brong, dubbed "Alkali Bill," who claimed to have made the run in 3 hours and 40 minutes.

There were seven Montgomery brothers, but it was Earnest Alexander "Bob" (standing, left) and George (standing, right) who, as author Richard Lingenfelter writes, "started Death Valley's great gold boom . . . and opened the first big mines." George came looking for the Lost Breyfogle and claimed he had found it near the Nevada settlement of Pahrump. Bob followed him to the Death Valley region but got married and was living in Goldfield when he caught the Bullfrog gold fever in September 1904. (TF.)

Bob Montgomery prospected the Bullfrog district but found nothing of value. On his way back to Goldfield, he stopped at Howell's ranch and met a Shoshone man known as Hungry Johnny, later called Shoshone Johnny (pictured), who said he knew where gold might be. Montgomery gave him some store credit and location notices. When Montgomery returned three weeks later, Johnny took him to some claims where, at first, it seemed there was not much gold.

It took an old prospector to show Bob Montgomery where the gold was on Shoshone Johnny's claims. In February 1905, they found gold running at $16,000 a ton. Montgomery exclaimed, "I have struck it; the thing that I have dreamed about since I was 15 years old. I am fixed for life and nobody can take it away from me." The Montgomery Shoshone mine became the hub of the district, and a new town—Rhyolite—sprang up in its shadow.

Bob Montgomery's mine was hailed as the "Wonder of the West," and Rhyolite, named for the rose-colored rock in the region, overtook Bullfrog as the area's largest mining camp. By June 1905, the town contained nearly 3,000 hopefuls, 50 saloons, 16 restaurants, 19 lodging houses, plenty of brothels, and the *Rhyolite Herald* newspaper. In this 1908 image, the three-story Cook Bank building is prominent as the population was at its peak (between 8,000 and 10,000). (CNHS.)

The sale of Bob Montgomery's mine to Charles Schwab, president of Pennsylvania's Bethlehem Steel, made Rhyolite boom into what some believed would become the "Chicago of the West." "Fussy silk and satin dresses were worn on the streets—dirt streets with no sidewalks," wrote Elizabeth Clemens, who was married to Earle R. Clemens, the editor of the *Rhyolite Herald*. Allan Eugene Holt, who took many photographs during the boom and marked them with his "A.E. Holt" signature, set up his real estate offices next to a store that likely serviced automobile, lighting, and heating needs. (WMC.)

Sen. William Stewart, who had been involved in the great silver camp of Panamint in the 1870s, was retired and back from Washington, DC, in 1905. He set up his law office in the new boomtown, helped with the sale of Bob Montgomery's mine, and was involved with a 1907 snake oil called Antifebrene, comprised of "healing" mud from hot springs near Grapevine Canyon in Death Valley. Stewart's law office sign is now in the collection of the Nevada Historical Society. (WMC.)

This photograph, taken in late 1905 or early 1906 on Colorado Street, shows the Wild West nature of Rhyolite. A stagecoach rumbles down the street. To the left is a barber pole and the Curtis Mann office building, where mining stocks and supplies were sold. On the corner at left is the Ice Palace saloon. At the far end of the street on the right is where the town jail would

later be built, and across from it were several houses of ill repute known as brothels or cribs. This was the boom period when four stages arrived daily, real estate was selling for $3,000 per lot, and buildings were going up to replace tents and small wooden shacks.

The Gibraltar was one of the rich mines in the Rhyolite district. The settlement on the edge of Death Valley was growing and becoming civilized. Soon, three railroads would serve the town: copper king William Andrews Clark's Las Vegas & Tonopah, borax king Francis Smith's Tonopah & Tidewater, and the Bullfrog Goldfield Railroad.

In Death Valley proper, there were only two real gold mines, and one of them belonged to Jack Keane and Domingo Etcharren. It was discovered in April 1904, and a small rush started, only to be outdone a few months later by Shorty Harris's strike at Bullfrog. The Keane Wonder Mine (pictured) had been nearly forgotten until 1906, when an investor bought out Keane and Etcharren and made the Keane Wonder a solid and steady mining proposition.

Another person who wielded considerable influence in the Death Valley gold rush was Bob Montgomery's wife, Winnie Aubrey. She insisted that Bob's Montgomery-Shoshone mine sell for at least $2 million, as they should each have $1 million. This is what brought steel king Charley Schwab to the desert to make the deal. When Bob bought 23 claims in the Panamints, Winnie named them Skidoo after the popular phrase "23 Skidoo," meaning "get out!" (TF.)

In 1906, as two prospectors were crossing the Panamints in a fog, they went searching for water and instead found gold. The result was the discovery of the district Winnie Aubrey Montgomery named Skidoo. This was Death Valley's second gold mine with an abundance of exposed gold ore. The Skidoo Mill had 10 heavy stamps or crushers to break the hard rock in order to extract the gold. According to historian Richard Lingenfelter, this enabled Bob and Winnie to do "what no Bullfrog mine ever did: it actually became a dividend payer!"

HOYT BOYS' LEASE THE SENSATION OF THE WEEK

Nothing less than phenomenal are the results that are being realized upon the Hoyt Brothers' lease on a portion of the Skidoo ground. Specimen ore, which is growing richer with every shot, is being taken out of the shaft is such quantities as to cause one to believe that they are opening a second Mohawk!

They are piling and carefully guarding their ore which is so rich that scarcely a single piece of quartz can be found that does not show free gold to the naked eye. As strong as this statment may seem, the News knows it to be a fact, from personal investigation.

"If it showed up any richer, we couldn't hold ourselves," said the Hoyt boys when a News representative visited the place where these fortunate leasers are rapidly sinking on rich ore. "If we don't see big nuggets of gold in every piece of rock, we don't consider it up to the average. We started upon a stringer of six inches of rich gold quartz and now at twelve feet the vein has widened to three feet with a continual gain in values at every shot. We have not had assays made of our ore—we can see so much of the gold that we know the ore is very rich and we haven't time to bother with assays. We are making every hour count in development. The way our ledge is developing is a great surprise to us. We knew it was rich at the surface but we had no idea that the ore shoot would increase in width and values as it is now doing. It is certainly a wonder now and the indications are that it will

be still greater as depth is gained."

Work on the Hoyt lease is being pushed with all possible speed. The point where work is being done is on the ridge just south of the townsite, and their shots resounding through the town proclaim that another step downward into golden riches has been taken. The ledge can be easily traced down into a steep gulch and presents an ideal tunneling proposition as soon as the character of the ore body has been demonstrated.

At the mouth of the shaft where sinking is now in progress there are two dumps, one large and the other small, proportionately, representing the week's work. The large dump is the higher grade ore, carefully sorted, and the smaller pile is the lower grade.

Messrs. Hoyt are very enthusiastic over their lucky strike and are confident of making a fortune out of their Skidoo lease.

An article in the *Skidoo News* described the 1906 discovery by Charles S. and Carl H. Hoyt, who were leasing land to mine from Bob Montgomery's Skidoo Mines Company. The Hoyt brothers are quoted as saying, "If we don't see big nuggets of gold in every piece of rock, we don't consider it up to the average!" The brothers are emblematic of Death Valley's gold rush and the confidence that they will indeed "make a fortune out of their Skidoo lease." (TF.)

Brothers Charles S. (left) and Carl H. Hoyt are pictured in front of their Skidoo cabin, crushing rock and preparing to do an assay on the ore. The mortar and scales used for weighing the rock are visible on the table. These portable assay scales were very practical and could be used on location to get a sense of the ore's value before going to get a more sophisticated opinion at a bank or assay office. (TF.)

In the only known photograph of the interior of the Skidoo Mill, Charles (left) and Carl Hoyt show off a sheet of gold. Discoveries like those of the Hoyt brothers helped to promote Skidoo, and by December 1906, the town had a newspaper, and rumors abounded that the town would grow bigger than Rhyolite. There was a telegraph line from Skidoo to Rhyolite and a road that went through Death Valley's sand dunes near where the Jayhawkers burned their wagons. (TF.)

Here, Charles (left) and Carl Hoyt are in their cabin toasting to their good fortune with an onlooking unidentified friend (center). The details of this interior show a lot about mining life in Skidoo. The walls are likely lined with newspaper, which was a common insulation in those days. The wood-burning stove and several frying pans are also visible. (TF.)

Charles (right) and Carl Hoyt pose in front of the door to their cabin with their feet on what appears to be a Native American blanket. A short version of Charles's name ("CHAS") is shown in the pattern on his hobnail boots. The good times did not last, as the financial panic of 1907 brought an end to most of the small mines in Death Valley, including that of the Hoyt brothers. (TF.)

When the town of Skidoo closed up, Charles and Carl Hoyt knew they had to move. This photograph is not specifically identified as featuring Skidoo, but the back of it is labeled "Moving Day." Although it was rarely captured on film, this type of scene would have been familiar to all who had to leave when mines played out, investment money dried up, and towns went bust. This is how Death Valley's ghost towns were born. Many items were left behind, including the table shown behind the rear wheel. (TF.)

Montillion Murray Beatty moved to this ranch east of Death Valley in the spring of 1896. He married a Shoshone woman, and they had two children, who are shown in this image. Beatty's ranch was a welcoming place for anyone passing through on their journeys. He lived there peacefully until the 1904 Bullfrog strike, when he sold his ranch, named the nearby settlement Beatty, and started a new ranch in Death Valley at a place called Cow Creek.

In 1904, the town of Beatty was only seven miles from the booming Bullfrog district. While Bullfrog had gold, Beatty had water. Because it was near the headwaters for the great Amargosa River, it was the perfect location for a supply camp. This promotional advertisement for Beatty touts its population, which grew from 0 to 1,500 in six months; the construction of a "modern" hotel; and the approaching railroads. The "E.A. Montgomery" referred to near the bottom of the advertisement is none other than Bob Montgomery. (TF.)

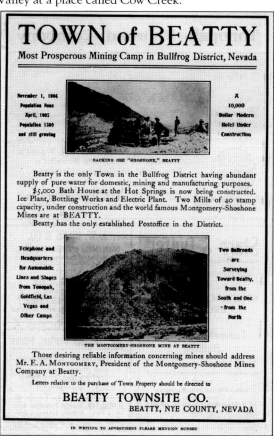

In this letter, prospector James Hargrove writes from his tent in Beatty on March 18, 1907, just six months before the Panic of 1907. Hargrove is full of optimism as he tells his friend in Montana that he has learned all about formations and "numerous other valuable points in mining and prospecting, and when I buy me a jackass [and] start out, how I can fail is more than I can see." (TF.)

These prospectors have their burros packed and are ready to leave Beatty to find some gold. It appears that the little girl was with the young and sporty man at left. The two prospectors are to the right, and the person standing behind the burros appears to be a businessman of some kind, possibly a grubstaker of the prospectors. Beatty was clearly growing, as a café, a saloon (the Northern), and a general store are visible in the background behind the men. (TEC.)

BARBER SHOP OF MARKS + HANCOCK. BEATTY

In the earliest days of Beatty, the businesses were located in tents, including the Marks & Hancock Barbershop shown here in 1905. The fancy, newly painted pole must have been red, white, and blue, as suggested by the stars in the center. The shop was a news depot as well, carrying the latest magazines. Within a year, businesses were springing up in Beatty, including an ice company and a water company. By 1906, water could be supplied for 20¢ per barrel, replacing the 1905 price of $2.50 per barrel. Saloons could even offer ice-cold beer, which could sometimes be as much of an attraction as finding gold. Due to Beatty's location and water supply, it lasted well past 1910, ensuring its survivability. It became, as author Robert McCracken writes, "the economic and social center for a very large geographical area." It remains an important desert outpost to this day. (TEC.)

Bob Montgomery supported the establishment of Beatty by building one of the finest hotels south of Reno and calling it the Montgomery. Here, the hotel is fully decked out for Railroad Days, celebrated on October 22 and 23, 1906, to herald the arrival of William Andrews Clark's Las Vegas & Tonopah Railroad. Montgomery sold the hotel, and it was moved to the nearby boom camp of Pioneer in 1909, where, sadly, it later burned down. (TEC.)

Although the Railroad Days celebrations were not held until October 22 and 23, 1906, The Las Vegas & Tonopah Railroad arrived in Beatty on October 18. It connected the Bullfrog district south to William Andrews Clark's San Pedro, Los Angeles & Salt Lake Railroad line at Las Vegas. April 25, 1907, marked the arrival of the Bullfrog Goldfield Railroad, which ran to Goldfield in the north. The last railroad to arrive was Borax Smith's Tonopah & Tidewater Railroad, which arrived in Beatty on October 27, 1907. The Tonopah & Tidewater lasted longer than any other railroad in the region, with its last train rumbling over the line in June 1940.

Six

POCKET MINERS

My Adventures with Your Money was the perfect title for the autobiography of George Graham Rice, whose real name was Jacob Herzig, as he was responsible for some of the biggest mining swindles in the United States. Rice had a history of conning people, and he brought his talents to Bullfrog, where he hyped the gold rush and sold mining stock totally out of proportion to its actual value. He also promoted a little mining camp on the edge of Death Valley called Greenwater, where some copper was found. Rice helped promote stock sales there to the tune of nearly $30 million beyond the actual $5,000 value. While the term "pocket miner" technically refers to miners who work a "pocket" of valuable mineral in the ground, for this chapter, it refers to those who were experts at mining the pockets of would-be investors and other suckers.

Rice was not the only pocket miner in the area. Death Valley's remote location was a great place for tales of hidden riches and mystery mines. No one knew this more than Walter Edward Scott, better known as Death Valley Scotty. He went to great lengths to convince investors and the press that his secret Death Valley mine was as rich as he said it was. Ultimately, he found one man who would not only invest but also enable his ruse.

One of the last to mine the pockets of gullible Death Valley investors was Charles Courtney Julian, better known as C.C. Julian. He got his start in Los Angeles promoting oil wells and his Julian Petroleum Company, popularly called the Julian Pete Oil Company. When that failed, he turned his sights to Death Valley and its hidden treasures. Hiring a special train and conducting a little mining, Julian caused a minor sensation in Los Angeles using Death Valley's mystique. The lengths to which Julian would go to convince potential investors that his mine was legitimate were impressive.

In the following years, the growth of modern media, the construction of paved roads, and Death Valley's inclusion in the National Park System would make it more difficult for people to pull off a con. But the lure and mystery of this valley named Death remains even today.

Once Bob Montgomery's Montgomery-Shoshone Mine went into operation, it was touted as "richer than the mines of King Solomon." Every mine around it was also certain to be rich. The promotion of the Bullfrog mines blew everything out of proportion. Papers like the *Bullfrog Miner* helped boost the town with articles and ads selling stock and inflating the values of everything in the region. (WMC.)

No one knew how to promote worthless stock better than George Graham Rice, whose birth name was Jacob Simon Herzig. He had spent time in reform school in New York, continued to gamble and forge checks on his father's account, went to New York State Prison, changed his name to George Graham Rice, moved to New Orleans, then left town and headed west to Tonopah and Goldfield. Later, the US Post Office accused him of fraud.

George Graham Rice opened an advertising agency in Goldfield and began promoting Rhyolite and mining stocks. The success of the Montgomery-Shoshone Mine gave Rice all the fuel he needed to make outlandish promises of wealth to the most hopeful of investors. With claims of vast fortunes being made and "the Greatest Gold Discovery of Modern Times," how could anyone resist? This stock is from the original Bullfrog mine, which was the first to ship gold ore from the district and one of the few to pay a dividend. (WMC.)

There were more than 200 Bullfrog mines, and nearly every one of them used the jumping critters in one way or another. They were used for logos, pendants, jewelry, cartoons, posters, and romantic postcards, such as this one from 1906. "Paul and Virginia," who were clearly a happy couple from the "Metropolis of the Bullfrog Mining District," served as a lighthearted invitation to come to Bullfrog and find happiness and wealth. (WMC.)

If Bullfrog was not enough, there was Greenwater. When copper was discovered with a spring running over green rocks, the stage was set for one of America's largest mining swindles. Here, men using a windlass haul ore to the surface of the Copper Queen Mine in 1906. Because surface showings were rich, this small copper mine created high hopes and attracted big money (from financiers such as Charles Schwab), raising the confidence of swindlers and naive investors.

Arthur Kunze brought in the first investors, creating excitement and spurring the building of a town. Standing at center is Ralph Jacobus "Dad" Fairbanks, who hauled water to the dry boomtown and sold it for no less than 10¢ per gallon ($3.15 today). Fairbanks eventually moved his family to Greenwater. When the town went bust, Fairbanks hung on until 1909, when he moved to the springs at Shoshone and took some of the buildings with him.

The little magazine called the *Death Valley Chuckwalla* exclaimed that one could do with it as they "damn well pleased, as soon as you paid for it!" It was designed to lure investors with its spectacular headlines and hype. After its 1907 debut, it found its way around the country, helping to sell stock in Greenwater mines. Years later, publisher Carl Glasscock admitted, "I'm a little ashamed now as I look over the vainglorious announcements and arrogant assertions in its pages." (TEC.)

Carl Glasscock was the cofounder and editor of *Death Valley Chuckwalla*. The *Chuckwalla* helped sell $30 million in stock for Greenwater mines while there was less than $5,000 worth of copper in the ground. Mollie Wright, sister-in-law of Arthur Kunze, brought investors to the camp. Pictured here are, from left to right, David Eldredge, printer Harry Glasscock (seated on the ground), *Chuckwalla* cofounder Curt Kunze (seated), Carl Glasscock (standing behind Kunze), and Wright. In 1909, Eldredge perished in Death Valley, and his body was found by Ralph "Dad" Fairbanks.

Goldfield Nev.

In *Death Valley Men*, author Bourke Lee writes, "All Death Valley is divided into three parts: Death Valley, itself; the Death Valley mountain ranges; and Death Valley Scotty. The greatest of these is Death Valley Scotty." This photograph of Scotty was taken in Goldfield around 1906. He was born Walter Edward Scott in Cynthiana, Kentucky, in 1872 and ran away at age 16, joining his brothers in Nevada. He learned how to be a world-class bronco buster and trick rider, taking his skills to New York and working in Buffalo Bill's Wild West Show. In 1902, Scotty was late for work, and Buffalo Bill docked his pay, so Scotty quit and pursued other plans. He took a couple of gold nuggets his wife, Josephine, had gotten on a mine tour in Colorado and met with Julian Gerard, the vice president of the Knickerbocker Trust Company. Gerard had a fascination with the West, so when Scotty told him the gold came from his secret mine in the "sinister reaches of Death Valley," Gerard was hooked. He gave Scotty $1,500 to work the mine for a half interest in all he found.

Death Valley Scotty left New York, "heading for the sands of Death Valley," as he said. Once in California, he wrote letters to Gerard extolling the incredible value of the riches "in sight" and the horrible Death Valley conditions he had to endure to get them. It worked to keep Gerard sending money yet staying far away. Scotty also managed to attract the attention of newspaper writers willing to help fabricate his larger-than-life image and perpetuate the mystery of his mine. (TF.)

'SCOTTY' PURCHASES AUTOMOBILE; EXPECTS TO BREAK MANY RECORDS

DEATH VALLEY MYSTERY IN HIS NEW CAR

This is a rare image of Death Valley Scotty—who is seated at far right along with his wife, Josephine (nicknamed "Jack")—in the tent saloon at Stovepipe Wells, which was named for a spring marked by a stovepipe. Scotty was building a reputation by buying rounds of drinks and bragging about his secret gold. The Death Valley gold rush helped fuel his story and instilled confidence in would-be investors. (CSH.)

In September 1905, Death Valley Scotty (kneeling at far left) took Azariah Y. Pearl into the backcountry of Death Valley to show him his secret mine. Scotty even let Pearl take a camera. Scotty brought along his brother Bill and two scouts. They headed to Scotty's camp in Death Valley, Camp Holdout, where his secret mine was located. Of course, they never made it to the mine, but Scotty showed Pearl various mining claims belonging to a half-Cherokee prospector

named Bill Keyes. As author Richard Lingenfelter points out, if there was one thing more important to Scotty than keeping himself in the headlines of the papers, it was "keeping the money coming in." Scotty knew that convincing Pearl of the value of his secret mine could lead to more investors and more money in Scotty's pocket. (CSL.)

On his 1905 trip with A.Y. Pearl, Death Valley Scotty had some of his men stage a bit of gunplay to make Pearl think that robbers were after them, so they headed back to the desert town of Barstow. Here, the group is stopped, likely for lunch. This was not the only time Scotty staged a shoot-out to scare and distract would-be investors. A year later, he took some Eastern investors

into Wingate Pass in Death Valley and had his friend Billy Keyes waiting in the rocks above the pass. As the group came into the pass, Keyes was to fire into the air to stage an ambush. But he hit Scotty's brother Warner, and Scotty called off the shoot-out. The famous "battle" of Wingate Pass exposed Scotty as a fraud. (CSL.)

Following the so-called Battle of Wingate Pass, the press turned on Death Valley Scotty, as a *San Bernardino Sun* editor wrote, "The Scotty boom has been as thoroughly punctured as that of any faking pretender who has flashed in and out of public view. He is a false alarm, a penny-ante bad man." Scotty had an impressive four-year run, but for the next decade, he would live in solitude in his desert cabin while keeping some of the old flames alive. (CSH.)

Death Valley Scotty fell hard in 1906, because just a year earlier, he had been at his peak. A mining promoter paid for a promotional record-breaking run of the Coyote Special train from Los Angeles to Chicago, carrying Death Valley Scotty and his wife, Jack. This marked the arrival of Death Valley Scotty on the national scene, promoting Death Valley and reconnecting Scotty to wealthy Chicago businessman Albert M. Johnson, who had invested in Scotty in 1904. This grand and highly publicized railroad run convinced Johnson and others that Scotty's claims were true. (TF.)

Ever hear of "Scotty" And his record-breaking ride?

This is a picture of him, and of his dog with the $1,000 collar.

The story? Briefly this— Walter Scott, the Death Valley gold miner, made the trip from Los Angeles to Chicago last summer on a special train over the Santa Fe in less than 45 hours. That whirlwind train cost him more than $6,000. It was the fastest long-distance run over mountains and plains ever made on any American railway. It demonstrated beyond dispute that the Santa Fe track, equipment and employes are of the dependable kind.

Probably you wouldn't care to ride so fast. You prefer the luxurious

California Limited

at $110 the round trip from Chicago. In service the year 'round between Chicago, Los Angeles and San Francisco. Now semi-weekly; daily, beginning early in November.

Santa Fe All the Way

It was the Coyote Special that convinced A.Y. Pearl and Albert Johnson that Death Valley Scotty's mines were real. This resulted in Scotty's two big gunplay charades, which cost him his reputation in the following years. Johnson suffered from a broken back due to a railroad accident. He was fascinated with Scotty (standing) as a Wild West character and loved his stories. Johnson (seated) began making trips to Death Valley—the weather helped his back, and he enjoyed spending time with Scotty. (TF.)

Albert Johnson bought land in the valley's north end and, in 1925, started construction on an elaborate Spanish-styled hacienda that Death Valley Scotty called his "castle." The name stuck, and it became Scotty's Castle. When the Great Depression depleted Johnson's money, the castle went unfinished, but it still became Death Valley's most famous attraction. Nearly 20 years after Scotty died in 1954, the castle was purchased by the National Park Service, and Scotty's colorful story is kept alive to this day.

Charles Courtney Julian (center) was described by Richard Lingenfelter as "a flamboyant, hayseed huckster with a pungent but disarmingly folksy style—a sly fox, thin, wiry and shrewd." Julian got his business start in the oil industry of Southern California. With four legitimate gushers he discovered, he formed Julian Petroleum Company and attracted investors. However, he soon discovered that working the pockets of investors was easier than discovering more oil.

Leadfield, Julian's Lead Mine, and Julian Himself

Setting and Promoter of Spectacular Stock-Selling Campaign
View of Leadfield in foreground. Its main street is "Julian avenue." The tents are those of miners employed by Julian. The Western Lead mine include most of the mountains in the background and the workings consist of a 100-foot horizontal drift and cross-cuts starting at the arrow. Inset is C. C. Julian examining a piece of lead ore.

C.C. Julian got out of the petroleum business when he found no more oil, but he roared back onto the scene in February 1926 with his Western Lead Mines Company. Death Valley contained hidden treasures, and he had discovered them. The mine was named Jazz Baby, and he "honestly believed" he had his arms around a "Hundred Million Dollar Silver-Lead Mine!" This image shows a newspaper article and one of the ads Julian used to attract gullible investors. (TF.)

Before C.C. Julian could sell any stock, he had to create the illusion of a mine and boom. He bought 14 worthless claims and hired engineers and geologists to create false claims about the wealth of the properties. He also graded a road into Titus Canyon from the east, not far from the railroad town of Beatty. He dug mine tunnels and built a small town in the canyon, which he called Leadfield. The trap was set.

On Saturday, March 3, 1926, C.C. Julian brought a train from Los Angeles with 340 potential investors on board. When they arrived in Beatty on Sunday morning, 90 private cars hired by Julian drove the visitors the 22 miles to Leadfield. The road gave them a thrill ride (which it still is today). When they arrived in Leadfield, there were about 800 enthusiasts who had come down from Tonopah and Goldfield. It appeared that the camp was booming.

The visitors who came to Leadfield on March 3, 1926, were treated to what seemed to be a booming mining town. There were speeches by experts who vouched for the wealth of the discoveries, a free lunch was served, a band played jazz, and Nevada's lieutenant governor inspired the crowd. The guests were given tours of the tunnels of the Jazz Baby mine to see it in all its glory. It worked. The next day, stock went up to $3.30, setting a record for trading on the Los Angeles Stock Exchange. The below image from the Leadfield event sums up the exhaustion people felt after dealing with the con artists who mined the pockets of gullible investors. An investigation into Western Lead Mines Company and C.C. Julian's dealings led to its downfall. Stocks plunged, and angry investors had to be stopped from rushing onto the exchange floor in Los Angeles. Julian tried to make a comeback, but when a former partner sued, he went to China, wrote his memoirs, and eventually committed suicide.

Seven

GOLDEN DREAMS

There was never enough gold in the ground to support the tales and rumors of Death Valley's vast treasures. As always, the people who made money were those who sold goods and services to others trying to get rich. Of course there were some prosperous mines, but they would all come to an end. Of all Death Valley's golden dreams, Rhyolite embodies them most dramatically.

It is relatively easy to find photographs of Rhyolite as it is today. But the life cycle of a Western ghost town is worth exploring. Rhyolite is, after all, one of the most famous ghost towns in the West, and its remains are arguably some of the most photographed ruins in America.

It was such a hopeful and energetic place, reaching a peak population of more than 7,000, yet it lasted only from 1905 to about 1913. The Panic of 1907 (a forerunner of the 1929 crash) ended the Death Valley gold rush and doomed Rhyolite. Investment money dried up, stocks became worthless, and the boom went bust. When the Montgomery-Shoshone Mine closed in 1911, it was the nail in the coffin, and by 1912, whole city blocks had been deserted, and the last newspaper shut down. In 1913, the post office closed, and by 1916, the trains had stopped coming through, and the power, lights, and lines were gone.

With a sort of natural send-off, Death Valley obliged the town's demise by offering the hottest summer on record, with temperatures soaring to 134 on July 10, 1913, which is still the world record. What had been a desert metropolis with hopes and dreams of becoming a vibrant city built on gold was now in ruins.

Elizabeth Clemens, the wife of newspaper publisher Earle Clemens, used the pseudonym Betsy Ritter to write her book *Life in the Ghost City of Rhyolite, Nevada*. She summed up the experience of many: "Moving day from Rhyolite was a beautiful day in May, 1911. We were on our way out leaving memories, precious friends and an appreciation of worth-while aspirations packed away in wrappings of broken dreams . . . accompanied by a tinge of sadness . . . and the desert itself . . . was waving goodbye to two greenhorn seekers of gold who at last had admitted their defeat."

When publisher Earle Clemens came from Wyoming, he started the *Rhyolite Herald* newspaper in a tent. He wrote, "The 'rainbow chasers' were crowded into canvas lodging houses, partitioned with cheesecloth or burlap. They were fed at lunch counters and often stood in line for an hour or two waiting for their turn. There were several very cold days. Can you imagine a thousand or two, men, women and children huddled together in rag houses . . . depending entirely upon the mercies of the weather to keep from freezing when a cold snap came?" (TEC.)

"The town was built in a hurry," wrote Elizabeth Clemens, "but not as quickly as might be thought. Patience, planning and waiting, long tiresome overland hauling and huge capital expenditures were put into its building." But Rhyolite was booming. By June 1905, there were 2,500 people and, according to Richard Lingenfelter, on the streets, a "confusion of pack mules, burros, dogs, horses, autos, stagecoaches and freight wagons." Here, wood is being hauled in to replace the tents.

By 1906, wooden structures had replaced the tents used for housing in the early days of the boom camp. Elizabeth Clemens recalls that she and her family moved into "a two-room house paying $35 a month rent; two rooms nicely papered, but no closets, cupboards, shelves or built-in breakfast nook. But we owned a tiny cook stove, with an oven just big enough in which to get the tips of the toes of two frozen feet." (TEC.)

In this photograph looking down Rhyolite's main thoroughfare, Golden Street, the construction of a new building is visible in the lower left corner. Demands for running water were met with three pipelines, and an ice plant and bottling works were built, according to Elizabeth Clemens. She wrote, "The district has the advantage of electric power and lights . . . with a phone exchange upon the lines of the Southern Nevada Consolidated Telephone & Telegraph company." (HLM.)

119

By 1908, Rhyolite had two of the "finest three story stone and concrete office buildings in Nevada, the finest passenger depot in the State and one of the best public school buildings in the southern part of the commonwealth," bragged Elizabeth Clemens. (CNHS.)

According to Elizabeth Clemens, "The years 1906, 7, 8 and 9 were really the years of the greatest activity. Shipments of ore were sent out by mule teams and horse teams while the railroads were being built. During the busy days of 1907 a full-fledged stock exchange was opened. Prospectors, miners, stock brokers, drug clerks, gamblers, bankers newspapermen and home makers were just a sample of the 10,000 inhabitants who made up the population of the Metropolis of the Bullfrog district of Death Valley Days of 1905–1909." (CNHS.)

By October 1907, three railroads had arrived in Rhyolite. Here, a Bullfrog Goldfield Railroad train is on its way out of town. The Las Vegas & Tonopah, with its beautiful Spanish-style depot, was the first to arrive (in 1906), and the line belonging to borax king Francis Marion Smith, the Tonopah & Tidewater, arrived in 1907. (TEC.)

Rhyolite had its own Nob Hill with beautiful multiroom houses, fireplaces, and front porches. "They were," as Elizabeth Clemens wrote, "showplaces, with green lawns and flower gardens running up the water bills to the tune of ninety dollars a month!" Here, John Overbury (in a suit) poses with his driver and family outside his home. Overbury was one of the few who made his fortune in the Bullfrog district. (HLM.)

The first truly permanent structure built in Rhyolite was the Overbury Building on Golden Street. Constructed in just three weeks with locally quarried stone by stonemasons who were paid the exorbitant rate of $1 per hour (more than doctors made) and costing more than $50,000, it became home to the Southern Nevada Bank. John Overbury's dream was to have the largest building in Rhyolite, which he did for a while. (CNHS.)

THE JOHN S. COOK BANK

John Overbury's building was overshadowed by banker John S. Cook's magnificent three-story, four-floor office building made of concrete, steel, and glass. It was completed in January 1908 and cost over $90,000. Author Richard Lingenfelter describes its "fancy time locks for the vaults, Italian marble stairs, imported stained-glass windows, and Honduran mahogany trim." It was Rhyolite's pride and joy, symbolizing the town's permanence. (CNHS.)

On March 14, 1911, the Montgomery-Shoshone Mine finally shut down. At that time, there were fewer than 600 residents left in town. On April 8, 1911, Earle Clemens wrote the final editorial for his *Rhyolite Herald*: "The boom came, and the boom went. Fortunes were made and spent, towns were built and torn down again. And nothing has occurred to stem the fatal tide." Golden Street and the fine Overbury Building (left) were in ruins within a few short years. (CNHS.)

The banks had failed shortly after the Panic of 1907, and by 1916, the magnificent John S. Cook bank building was in ruins. As author Richard Lingenfelter describes, "The big steel vault doors were taken off and sold, the floor strewn with canceled checks and worthless stock certificates. The desert metropolis had become a 'graveyard of blasted hopes.'" (WMC.)

A visitor to the ghost town wrote on a postcard: "Abandoned shacks in Rhyolite, Nev. Called the 'Ghost City.' During the gold rush in 1906-7 Rhyolite had a population of 5000. It is said that people had to elbow their way through the streets. Many substantial buildings were erected, some of which still stand dismantled of everything portable. At the beginning of 1920 Rhyolite is deserted with the exception of four persons." (WMC.)

In 1906, Earle Clemens wrote, "It is a pleasure to know that the world holds such men as William Andrews Clark who are willing to risk millions of dollars in the opening up of new territory to commerce, depending solely upon the resources of new and undeveloped fields to furnish business for the project." By 1918, Clark's Las Vegas & Tonopah Railroad was out of business and sold for scrap. His stylish depot was one of the ghost town's main attractions, as it is to this day. (WMC.)

Elizabeth Clemens writes, "Gold was the magic word which brought Rhyolite into existence." It was the reason why people left their homes to go to the desert and, in one case, make their new home out of bottles—Tom Kelly used 50,000 beer and liquor bottles to make his. When Paramount Pictures used the town as a location for a film called *The Air Mail* in 1925, the studio worked with the Beatty Improvement Association to have the house restored. (WMC.)

After 1913, most of the wood was hauled off, and the town of Rhyolite was left to decay. Few buildings stand today, but it remains one of the West's most visited ghost towns. The final words of Earle Clemens's last editorial sum up the feelings many had about the era: "May prosperity again reign in Rhyolite—the prettiest, coziest, mining town in the great American desert; a town blessed with ambitious, hopeful courageous people and with a climate second to none on earth. Goodbye, dear old Rhyolite." (CNHS.)

Bill Murphy and his brothers Dan and Tom came from Montana looking not only for gold but for something else. This photograph shows, from left to right, Dan, Tom, an unidentified man holding a chunk of ore, and Bill. In the film *Death Valley Memories*, Willis Keyes, the son of Billy Keyes, Death Valley Scotty's gunslinger in the Battle of Wingate Pass, states, "Death Valley was really the last holdout of the old West." Reliving the Wild West was just as appealing as the quest for gold, and thousands of hopefuls threw themselves at an indifferent landscape that led to some returning in caskets, left others broke, and provided a few with work and fewer still with success. The Murphy brothers never got rich, but they worked in the Tonopah mines. Years later, Bill was called "Big Bill," as he died rescuing others in Tonopah's tragic Belmont Mine fire. The brothers had been lured by the Wild West and gold at a time when just about anyone with a donkey, a pick, some beans, and a shovel could try their hand at making a strike in the Death Valley gold rush and shouting "Eureka! I have found it!" (Murphy family.)

BIBLIOGRAPHY

Burdick, Arthur J. *The Mystic Mid-Region.* New York: G.P. Putnam's Sons, 1904.
Caruthers, William. *Loafing Along Death Valley Trails.* Ontario, CA: Death Valley Publishing, 1951.
Cronkhite, Daniel. *Death Valley's Victims.* Morongo Valley, CA: Sagebrush Press, 1981.
Earl, Phillip I. *This Was Nevada.* Reno: Nevada Historical Society, 1986.
Ellenbecker, John G. *The Jayhawkers of Death Valley.* Marysville, KS: Self-published, 1938.
Elliott, Russell R. *Nevada's 20th Century Mining Boom.* Reno: University of Nevada Press, 1966.
Faye, Ted. *Death Valley Memories.* Los Angeles, CA: Gold Creek Films, 1994.
———. *The Twenty Mule Team of Death Valley.* Charleston, SC: Arcadia Publishing, 2012.
Glasscock, Carl B. *Here's Death Valley.* Indianapolis and New York: Bobbs-Merill Co., 1940.
Gold Creek Films. www.goldcreekfilms.com.
Johnson, Jean and LeRoy. *Escape from Death Valley.* Reno and Las Vegas: University of Nevada Press, 1987.
Johnson, Jean and LeRoy, with Ted Faye and Robert Ryan. *John Rogers, Death Valley's Unsung Hero of 1849.* Death Valley, CA: Death Valley '49ers, Inc., 1999.
Koenig, George. *Beyond This Place There be Dragons.* Glendale, CA: The Arthur H. Clark Company, 1984.
Lakes, Arthur. *Prospecting for Gold & Silver.* Scranton, PA: Colliery Engineer Co., 1895.
Latta, Frank F. *Death Valley '49ers.* Santa Cruz, CA. Bear State Book, 1979.
Lee, Bourke. *Death Valley.* New York: The MacMillan Company, 1931.
Lingenfelter, Richard E. *Death Valley and the Amargosa, A Land of Illusion.* Berkeley and Los Angeles: University of California Press, 1986.
Manly, William Lewis. *Death Valley in '49.* San Jose, CA: Pacific Tree & Vine, 1894.
McCoy, Suzy. *Rebecca's Walk Through Time, A Rhyolite Story.* Lake Grove, OR: Western Places, 2004.
McCracken, Robert. *Beatty: Frontier Oasis.* Tonopah, NV: Nye County Press, 1992.
Patera, Alan H., and David A. Wright. *Skidoo.* Lake Grove, OR: Western Places, 1999.
Pipkin, George C. *Pete Aguereberry.* Trona, CA: Murchison Publications, 1971.
Ritter, Betsy. *Life in the Ghost City of Rhyolite, Nevada.* Morongo Valley, CA: Sagebrush Press, 1982.
Stephens, L. Dow. *Life Sketches of a Jayhawker of '49.* New York: Sagwan Press, 1916.
Walker, Ardis Manly. *Death Valley & Manly, Symbols of Destiny.* San Bernardino, CA: Inland Printing, Inc., 1962.
Weight, Harold, and Lucile Weight. *Lost Mines of Death Valley.* Twentynine Palms, CA: Calico Press, 1953.
———. *Wm. B. Rood.* Twentynine Palms, CA: Calico Press, 1959.
Wynn, Marcia Rittenhouse. *Desert Bonanza.* Glendale, CA: The Arthur H. Clark Co., 1963.
Zanjani, Sally. *Goldfield, The Last Great Gold Rush on the Western Frontier.* Athens: Swallow Press/Ohio University Press, 1992.
———. *A Mine of Her Own.* Lincoln and London: University of Nebraska Press, 1997.

DISCOVER THOUSANDS OF LOCAL HISTORY BOOKS
FEATURING MILLIONS OF VINTAGE IMAGES

Arcadia Publishing, the leading local history publisher in the United States, is committed to making history accessible and meaningful through publishing books that celebrate and preserve the heritage of America's people and places.

Find more books like this at
www.arcadiapublishing.com

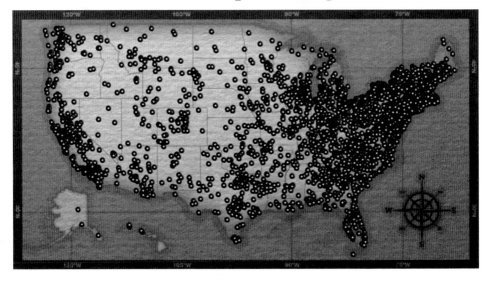

Search for your hometown history, your old stomping grounds, and even your favorite sports team.